A Level
Salters Advanced
Chemistry
for OCR

Year 1 and AS

David Goodfellow • Mark Gale

OXFORD
UNIVERSITY PRESS

OXFORD
UNIVERSITY PRESS

Great Clarendon Street, Oxford, OX2 6DP, United Kingdom

Oxford University Press is a department of the University of Oxford. It furthers the University's objective of excellence in research, scholarship, and education by publishing worldwide. Oxford is a registered trade mark of Oxford University Press in the UK and in certain other countries

© David Goodfellow and Mark Gale

The moral rights of the authors have been asserted

First published in 2016

British Library Cataloguing in Publication Data
Data available

978-0-19-833291-6

10 9 8 7 6 5 4 3 2 1

Paper used in the production of this book is a natural, recyclable product made from wood grown in sustainable forests. The manufacturing process conforms to the environmental regulations of the country of origin.

Printed in Great Britain by Bell and Bain Ltd, Glasgow

Artwork by Q2A Media

Although we have made every effort to trace and contact all copyright holders before publication this has not been possible in all cases. If notified, the publisher will rectify any errors or omissions at the earliest opportunity.

Links to third party websites are provided by Oxford in good faith and for information only. Oxford disclaims any responsibility for the materials contained in any third party website referenced in this work.

MIX
Paper from
responsible sources
FSC
www.fsc.org   FSC® C007785

AS/A Level course structure

This book has been written to support students studying for OCR AS Chemistry B (Salters) and for students in their first year of studying for OCR A Level Chemistry B (Salters). It covers the AS content from the specification, the content of which will also be examined at A Level. The content is arranged by chemical idea. The content covered is shown in the contents list, which also shows you the page numbers for the main topics within each chapter. If you are studying for OCR AS Chemistry B (Salters), you will only need to know the content in the shaded box.

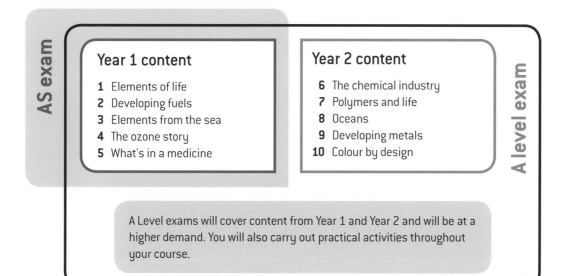

AS exam

Year 1 content

1 Elements of life
2 Developing fuels
3 Elements from the sea
4 The ozone story
5 What's in a medicine

Year 2 content

6 The chemical industry
7 Polymers and life
8 Oceans
9 Developing metals
10 Colour by design

A level exam

A Level exams will cover content from Year 1 and Year 2 and will be at a higher demand. You will also carry out practical activities throughout your course.

How to use this book

This book contains many different features. Each feature is designed to support and develop the skills you will need for your examinations, as well as foster and stimulate your interest in chemistry.

This book is structured by chemical idea.

Worked example
Step-by-step worked solutions.

Common misconception
Common student misunderstandings clarified.

Challenge
Familiar concepts in an unfamiliar context.

Maths skill
A focus on maths skills.

Model answers
Sample answers to exam-style questions.

Summary Questions

1   These are short questions at the end of each topic.

2   They test your understanding of the topic and allow you to apply the knowledge and skills you have acquired.

3   The questions are ramped in order of difficulty. Lower-demand questions have a paler background, with the higher-demand questions having a darker background. Try to attempt every question you can, to help you achieve your best in the exams.

Specification references
→   At the beginning of each topic there is a specification reference to allow you to monitor your progress.

Key term
Pulls out key terms for quick reference.

Revision tips
Revision tips contain prompts to help you with your understanding and revision.

Synoptic link
These highlight the key areas where topics relate to each other. As you go through your course, knowing how to link different areas of chemistry together becomes increasingly important. Many exam questions, particularly at A Level, will require you to bring together your knowledge from different areas.

Chapter 1 Practice questions

1   What mass of sodium hydroxide, NaOH is needed to form 200 cm³ of a 0.25 mol dm⁻³ solution?
A   2000 g
B   2 g
C   32 g
D   10 g
(1 mark)

2   Choose the phrase that correctly completes the following definition:
The relative isotopic mass is the mass of one atom of an isotope compared to....
A   The mass of a carbon-12 atom.
B   12 g of carbon-12 atoms.
C   one-twelfth of 12 g of carbon-12 atoms.
D   one-twelfth of the mass of a carbon-12 atom.
(1 mark)

3   1.40 g of ethene, C_2H_4, reacts with excess steam, H_2O, to form ethanol, C_2H_5OH. The equation for the reaction is:
$C_2H_4 + H_2O \rightarrow C_2H_5OH$
1.06 g of ethanol are formed in the reaction. The % yield of this process is given by the equation:

A $\frac{1.06}{1.40} \times 100$

B $\frac{\frac{1.06}{(1.40 \times 28)}}{46} \times 100$

C $\frac{\frac{1.06}{(1.40 \times 46)}}{28} \times 100$

D $\frac{1.40}{(1.06 \times 28)} \times 100$
(1 mark)

4   A compound that contains C, H, and O atoms only has the following % composition by mass: C: 40% O: 53%.
The relative molecular mass of the compound was found to be 60.
The molecular formula of the compound is:
A CH_2O

Practice questions including questions that cover practical and maths skills.

Experimental techniques
Specification reference: EL (b) (ii), EL (c) (ii), DF (a)

Weighing a solid
Masses of solids are normally measured on an electronic balance that records masses to two or three decimal places.

Weighing out a known mass of solid
1   Place a weighing bottle or similar container on the balance.
2   Add the required mass of solid.
3   Reweigh the weighing bottle + solid.
4   Transfer the solid to the reaction vessel or volumetric flask.
5   Reweigh the weighing bottle without the solid.

Heating to constant mass
In some experiments you need to measure the mass of a solid remaining after thermal decomposition. Thermal decomposition of hydrated salts can be carried out in a crucible supported on a pipe-clay triangle.

Weighing to constant mass ensures that decomposition is complete:
1   Weigh an empty crucible.
2   Add the hydrated salt and weigh the crucible + hydrated salt.
3   Heat the crucible strongly for several minutes.
4   Allow the crucible to cool and weigh the crucible and contents.
5   Heat strongly for a further minute and reweigh the crucible and contents again.
6   If the mass has changed, then repeat this last step until two successive weighings are identical.

▲ Figure 1 Heating a hydrated salt in a crucible.

Volumetric equipment
Pipettes and burettes are described as volumetric equipment, because they can be used to measure volumes of liquids accurately.

Dedicated experimental techniques pages, detailing all of the practical skills you need to know for your examinations.

Isotopes

Isotopes are atoms of the same element. They have the same atomic number (number of protons), but they have different mass numbers (different numbers of neutrons). Most elements exist naturally as a mixture of isotopes.

For example, there are two common isotopes of chlorine:

- $^{35}_{17}Cl$ has 17 protons and 18 neutrons.
- $^{37}_{17}Cl$ has 17 protons and 20 neutrons.

Relative atomic mass (A_r)

The **relative atomic mass** (A_r) is the average of the relative isotopic masses of the naturally occurring isotopes of an element, taking into account their abundances.

So the A_r value of an element tells you the average mass of an atom of that element compared to one-twelfth of an atom of ^{12}C. A_r values have no units.

> ### 🖩 Worked example: Calculating a relative atomic mass
>
> Silicon, Si, exists as three naturally occurring isotopes. The % abundances of the three isotopes are:
>
> $^{28}_{14}Si$: 92.2%, $^{29}_{14}Si$: 4.7%, $^{30}_{14}Si$: 3.1%. Calculate the relative atomic mass of silicon to 2 decimal places.
>
> **Step 1:** Multiply the % abundance by the relative mass of each isotope.
>
> **Step 2:** Divide by 100.
>
> **Step 3:** Round the answer to 2 decimal places.
>
> $$\frac{(92.2 \times 28)+(4.7 \times 29)+(3.1 \times 30)}{100} = 28.109 = 28.11 \text{ to 2 d.p.}$$

Relative formula mass (M_r)

The relative formula mass (M_r) is the sum of the relative atomic masses (A_r) for each atom in the formula. For a simple molecular compound, it is sometimes called the relative molecular mass. M_r values have no units.

Avogadro constant and the mole

The mole is the unit of **amount of substance**:

- To obtain 1 mole of a substance, weigh out the A_r or M_r exactly in grams.
- The Avogadro constant, $6.02 \times 10^{23} \text{mol}^{-1}$ (given the symbol N_A), is the number of particles in 1 mole of a substance.

Calculations involving amount of substance

Calculating amount of a substance (number of moles)

To work out the amount in moles of a substance, use:

$$\text{Amount in moles} = \frac{\text{mass in grams}}{A_r} \text{ for an atomic element}$$

$$\text{amount in moles} = \frac{\text{mass in grams}}{M_r} \text{ for a molecule or compound.}$$

Empirical and molecular formulae

Empirical formula

To find the empirical formula of a compound, you simply need to calculate the number of moles of each element in the compound.

> ### 🖩 Worked example: Calculating empirical formula
>
> A sample of aluminium oxide contains 0.310 g of aluminium and 0.276 g of oxygen. Calculate the empirical formula.
>
> | | Al (A_r = 27.0) | O (A_r = 16.0) |
> |---|---|---|
> | mass / g | 0.310 | 0.276 |
> | moles / mol | $\frac{0.310}{27.0} = 0.011\,48$ | $\frac{0.276}{16.0} = 0.017\,25$ |
> | ratio (divide each number by the smallest of the two numbers) | $\frac{0.01148}{0.01148} = 1.00$ | $\frac{0.01725}{0.01148} = 1.50$ (to 3 s.f.) |
>
> The ratio of Al to O is 1:1.5. This is not a whole number ratio; the simplest whole number ratio would be 2:3.
>
> The formula is Al_2O_3.
>
> If the masses are given as percentages (e.g. 52.9% Al and 47.1% O), then simply treat these numbers as masses (52.9 g and 47.1 g) to start the calculation.

Molecular formula

The molecular formula may be the same as the empirical formula, or it may be a whole number multiple of the empirical formula.

You can use the M_r value of the molecule to find the molecular formula.

> ### 🖩 Worked example: Molecular formula
>
> The empirical formula of a hydrocarbon molecule is CH_2. The M_r value is found to be 84.
>
> **Step 1:** M_r (CH_2) = 14.
>
> **Step 2:** Number of empirical formula units needed to give an M_r of 84 = $\frac{84}{14}$ = 6.
>
> **Step 3:** $CH_2 \times 6 = C_6H_{12}$.
>
> The molecular formula is C_6H_{12}.

Water of crystallisation

Some ionic salts are hydrated. They contain water of crystallisation, which can be driven off when the salt is heated.

Experimental data can be used to find the formula of the ionic salt.

> ### 🖩 Worked example: Finding the formula of a hydrated salt
>
> Hydrated cobalt(II) chloride has the formula $CoCl_2 \bullet \times H_2O$.
>
> 1.173 g of hydrated cobalt chloride is heated to drive off the water of crystallisation. The mass remaining is 0.641 g.
>
> **Step 1:** Calculate the mass of water driven off: 1.173 − 0.641 = 0.532 g.
>
> **Step 2:** Calculate the number of moles of water (M_r = 18): $\frac{0.532}{18} = 0.029\,56$.
>
> **Step 3:** Calculate the number of moles of anhydrous salt (M_r = 129.9): $\frac{0.641}{129.9} = 0.004\,93$.
>
> **Step 4:** Calculate the ratio $\frac{\text{moles water}}{\text{moles anhydrous salt}} : \frac{0.2956}{0.00493} = 5.996$.
>
> The formula of the hydrated salt = $CoCl_2 \bullet 6H_2O$.

Synoptic link

Relative atomic masses are calculated from the abundances of the different isotopes of an element as measured by a mass spectrometer. The use of a mass spectrometer is described in Topic 6.5, Mass spectrometry.

Revision tip

You should use the A_r values (to 1 d.p.) from the periodic table to calculate the M_r value of a compound.

Key terms

Avogadro constant: The number of particles in 1 mole of a substance. Unit: mol^{-1}.

Mole: An amount of substance containing the same number of particle as there are in 12 g of ^{12}C. Unit: mol.

Revision tip

If an element exists as molecules, e.g. oxygen (O_2), look carefully at the question to see whether you are being asked to calculate the number of moles of atoms or molecules. The relative masses of these particles are different!

Key term

Empirical formula: The simplest whole number ratio of atoms of each element in a compound.

Revision tip

If the ratio from a calculation is not a whole number (e.g. 1:1.67), then do not be tempted to round up to make the whole number ratio 1:2. Find a pair of whole numbers that match this ratio (in this case 3:5).

Key term

Molecular formula: the actual number of atoms of each element in a compound molecule.

Percentage composition

You can work out the percentage by mass of an element in a compound using the formula of the compound.

Worked example: Working out the percentage by mass of an element

Calculate the percentage by mass of nitrogen in ammonium sulfate, $(NH_4)_2SO_4$.

Step 1: $M_r (NH_4)_2SO_4 = 2 \times (14.0 + 1.0 + 1.0 + 1.0 + 1.0) + 32.1 + (4 \times 16.0) = 132.1$.

Step 2: Mass of nitrogen in 1 mole of $(NH_4)_2SO_4 = 14.0 + 14.0 = 28.0$.

$$\% \text{ by mass} = \frac{\text{mass of element in 1 mole}}{M_r \text{ of compound}} \times 100 = \frac{28.0}{132.1} \times 100 = 21.2\%$$

Summary questions

1 What is the M_r value of:
   a aluminium sulfate, $Al_2(SO_4)_3$ (1 mark)
   b naphthalene $C_{10}H_8$? (1 mark)
2 Calculate the % by mass of C in benzene, C_6H_6. (1 mark)

3 Relative isotopic masses have whole number values when measured to 2 decimal places, whereas relative atomic masses are often not whole numbers. Explain why. (2 marks)
4 The % by mass of C in CO_2 is 27%. However, 33% of the atoms in CO_2 are C. Explain the difference in these numbers. (2 marks)

5 Bromine and boron both have two naturally occurring isotopes:
   a The % abundance of the two isotopes of bromine are: ^{79}Br: 50.52%, ^{81}Br: 49.48%. Calculate the relative atomic mass of bromine to 2 d.p. (1 mark)
   b The two isotopes of boron are ^{10}B and ^{11}B. The relative atomic mass of boron is 10.8 to 1.d.p. Calculate the % abundance of each of the two B isotopes. (1 mark)
6 A compound containing C, H, and O only has a composition of 73.47% C and 10.20% H. The M_r is found to be 196. What is the molecular formula of the compound? (2 marks)

1.2 Balanced equations

Specification reference: EL(d)

Writing formulae

You can use the charges on ions to write the formulae of ionic substances.

> ### 🖩 Worked example: Writing formula of ionic compounds
>
> Write the formula of calcium chloride.
>
> **Step 1:** Charges on ions: Ca^{2+}, Cl^-.
>
> **Step 2:** To balance the charges, 2 Cl^- ions are needed for 1 Ca^{2+}.
>
> **Step 3:** The formula is $CaCl_2$.

> ### 🖩 Worked example: Writing the formula of more complex ionic compounds
>
> Write the formula of aluminium sulfate:
>
> **Step 1:** Charges on ions: Al^{3+}, SO_4^{2-}.
>
> **Step 2:** To balance the charges, 1½ SO_4^{2-} are needed for 1 Al^{3+}.
>
> **Step 3:** The simplest whole number ratio is $2Al^{3+} : 3SO_4^{2-}$.
>
> **Step 4:** The formula is written as $Al_2(SO_4)_3$.

Full chemical equations

Once the correct formulae have been written for the substances in a chemical equation, you can add balancing numbers to produce a full balanced equation.

> ### 🖩 Worked example: Writing a balanced equation
>
> A piece of calcium is added to water. The calcium reacts to form a solution of calcium hydroxide and hydrogen is given off.
>
> **Step 1:** Write the word equation: calcium + water $\rightarrow$ calcium hydroxide + hydrogen.
>
> **Step 2:** Put in the formulae: $Ca + H_2O \rightarrow Ca(OH)_2 + H_2$.
>
> **Step 3:** Balance the equation using large numbers before the formulae:
>
> $$Ca + 2H_2O \rightarrow Ca(OH)_2 + H_2.$$
>
> Check:
>
> | | Left | Right |
> |----|------|-------|
> | Ca | 1 | 1 |
> | O | 2 | 2 |
> | H | 4 | 4 |
>
> **Step 4:** Add the state symbols if the question asks for them:
>
> $$Ca(s) + 2H_2O(l) \rightarrow Ca(OH)_2(aq) + H_2(g).$$

Revision tip

You will need to learn some formulae of some common substances.

Water	H_2O
Carbon dioxide	CO_2
Hydrogen	H_2
Hydrochloric acid	HCl
Sulfuric acid	H_2SO_4

Revision tip

There are some formulae of common ions that you should also learn:

Zinc	Zn^{2+}		
		Hydroxide	OH^-
		Carbonate	CO_3^{2-}
		Sulfate	SO_4^{2-}
		Nitrate(V)	NO_3^-
		Ammonium	NH_4^+
Silver	Ag^+	Hydrogen carbonate	HCO_3^-

Revision tip

The charges of simple ions of elements in the s- and p-block of the periodic table follow a regular pattern. You can use this pattern to check the charges on simple ions such as Ba^{2+} and Cl^-.

Revision tip

You should use these state symbols: (s) = solid, (l) = liquid, (aq) = aqueous solution, (g) = gas.

Common misconception: Aqueous solutions and liquids

Both water and calcium hydroxide solution look like colourless liquids. Water is a pure liquid and is given the state symbol (l), but the calcium hydroxide solution contains a substance dissolved in water and (aq) is used.

Ionic equations

The term *ionic equation* is sometimes used to describe reactions involving ionic compounds in solution. Any ions that do not change during the reaction are left out, making it easier to see what has happened.

Worked example: Writing an ionic equation for a precipitation reaction

Silver nitrate solution is added to a solution containing chloride ions. A precipitate of silver chloride is formed. Write an ionic equation for the reaction, including state symbols.

Step 1: Write the formula of the precipitate: $AgCl$.

Step 2: Show the ions that combine to form this precipitate:

$$Ag^+ + Cl^- \longrightarrow AgCl.$$

Step 3: Add the state symbols:

$$Ag^+(aq) + Cl^-(aq) \longrightarrow AgCl(s).$$

Summary questions

1 What is the usual charge on an ion formed by an atom of these elements?
   a sulfur (Group 6)
   b rubidium (Group 1). (*2 marks*)

2 Write the formula of these ionic compounds:
   a barium hydroxide
   b aluminium oxide
   c sodium sulfate. (*3 marks*)

3 a A piece of calcium oxide reacts with hydrochloric acid to form water and a solution of calcium chloride. Write a full balanced equation for this process. (*2 marks*)
   b Barium chloride solution is added to a solution containing sulfate ions. A white precipitate of barium sulfate is formed. Write an ionic equation, including state symbols. (*2 marks*)

1.3 Using equations to work out reacting masses

Specification reference: EL (b) (i)

Calculating reacting masses

You can convert between masses and moles using these relationships:

$$\text{Moles} = \frac{\text{mass}}{A_r \text{ or } M_r} \qquad \text{Mass} = \text{moles} \times M_r \text{ (or } A_r)$$

 Worked example: Calculating reacting masses

What mass of aluminium is needed to completely react with 7.98 g of iron(III) oxide, Fe_2O_3?

Formulae and balancing numbers taken from the equation	Fe_2O_3	$2Al$
M_r or A_r value	159.6	27.0
Mass	7.98	$0.100 \times 27 = 2.70$ g
Amount in moles	$\frac{7.98}{159.6} = 0.0500$	0.100

Step 1: Convert the mass of Fe_2O_3 into moles using:

$$\text{moles} = \frac{\text{mass}}{A_r \text{ or } M_r}.$$

Step 2: Use the ratio of the balancing numbers from the equation to work out the moles of Al: $2 \times 0.0500 = 0.100$ mol.

Step 3: Convert the moles of Al into a mass using: mass = moles $\times M_r$ (or A_r).

Percentage yield

Reacting mass calculations allow you to calculate the expected yield of a reaction.

 Worked example: % yield

7.98 g of iron(III) oxide, Fe_2O_3, are reacted with excess aluminium. 4.62 g of iron are formed. Calculate the % yield (to 3 s.f.)

Step 1: Calculate the expected yield:

Formulae and balancing numbers taken from the equation	Fe_2O_3	$2Fe$
M_r or A_r value	159.6	55.8
Mass	7.98	$0.100 \times 55.8 = 5.58$ g
Amount in moles	$\frac{7.98}{159.6} = 0.0500$	0.100

Step 2: Compare the actual yield to the expected calculated yield:

$$\frac{4.62}{5.58} \times 100 = 82.645\ldots$$

Step 3: Give the answer to 3 significant figures: 82.6%.

Common misconception: Converting moles

Be careful when converting moles of product to mass of product (Step 3): do not try and use the A_r value of 2Al (54.0). The balancing number (2) has already been taken account in the second step.

Summary questions

1 When carbon is burnt in oxygen, it forms carbon dioxide $(C + O_2 \rightarrow CO_2)$. Calculate the mass of carbon dioxide formed from 10 g of carbon.
(1 mark)

2 Uranium hexafluoride (UF_6) is formed in a reaction between uranium and chlorine trifluoride (ClF_3): $U + 2ClF_3 \rightarrow UF_6 + Cl_2$. What mass of ClF_3 is needed to form 100 g of UF_6? *(3 marks)*

3 In a reaction to form UF_6, 1 kg of uranium is reacted with 1 kg of ClF_3. 1.36 kg of UF_6 are formed. Calculate the % yield of UF_6 in this reaction. *(4 marks)*

Calculating concentrations

Concentrations can be measured in grams per cubic decimetre ($g\,dm^{-3}$).

Example

A solution containing $40\,g$ of sodium hydroxide in $2\,dm^3$ of solution has a concentration of $\frac{40}{2} = 20g\,dm^{-3}$

A more common unit for concentration is *moles per cubic decimetre* ($mol\,dm^{-3}$).

1 mole of a substance, dissolved and then made up to a volume of $1\,dm^3$ of solution, has a concentration of $1\,mol\,dm^{-3}$.

> ### ⊞ Worked example: Converting from $g\,dm^{-3}$ to $mol\,dm^{-3}$
>
> A solution of sodium hydroxide, NaOH, has a concentration of $36\,g\,dm^{-3}$. What is the concentration in $mol\,dm^{-3}$?
>
> **Step 1:** Find the M_r value for NaOH: $M_r = 40$.
>
> **Step 2:** Convert $36\,g$ into moles: $\frac{36}{40} = 0.90\,mol$.
>
> The concentration is $0.90\,mol\,dm^{-3}$.

Moles = volume (in dm^3) × concentration (in $mol\,dm^{-3}$)

OR, if the volume is given in cm^3:

Moles = $\dfrac{\text{volume (in } cm^3)}{1000}$ × concentration (in $mol\,dm^{-3}$)

> ### ⊞ Worked example: Calculating concentrations from masses and volumes
>
> $5.85\,g$ of NaCl ($M_r = 58.5$) is dissolved and made up to $250\,cm^3$. What is the concentration in $mol\,dm^{-3}$?
>
> **Step 1:** Calculate the number of moles of NaCl: $\frac{5.85}{58.8} = 0.100\,mol$.
>
> **Step 2:** Calculate the volume in dm^3: $\frac{250}{1000} = 0.250\,dm^3$.
>
> **Step 3:** Rearrange the expression linking moles, volume and concentration:
>
> $$n = c \times v, \text{ so } c = \frac{n}{v}.$$
>
> **Step 4:** Substitute the numbers into this expression:
>
> $$c = \frac{0.100}{0.250} = 0.400\,mol\,dm^{-3}$$

Diluting solutions

Solutions are diluted by taking a small volume of the original solution and diluting it to a greater volume using distilled water.

> ### ⊞ Worked example
>
> $10.0\,cm^3$ of a $1.00\,mol\,dm^{-3}$ solution of HCl was transferred to a volumetric flask and made up to $250\,cm^3$ with water. What is the concentration of the diluted solution?
>
> **Step 1:** Calculate the dilution factor = $\frac{10.0}{250}$
>
> **Step 2:** Use this to calculate the new concentration of the diluted solution = $\frac{10.0}{250} \times 1.00$
>
> $= 0.0400\,mol\,dm^{-3}$

Revision tip

Volumes can be measured in dm^3 or cm^3. If volumes are given in cm^3, you will need to convert it to dm^3.

$1\,dm^3 = 1000\,cm^3$ or
$1\,cm^3 = \dfrac{1}{1000}\,dm^3$

So to convert cm^3 to dm^3, you need to divide by 1000.

Revision tip

You need to be able to rearrange the expressions linking moles, volume, and concentration. Practise doing this frequently!

Synoptic link

Details of the experimental method used to make up and dilute solutions can be found in Experimental Techniques.

Revision tip

You can calculate the new concentration of the diluted solution by using the dilution factor:

Dilution factor =
$\dfrac{\text{volume of original solution}}{\text{new volume of diluted solution}}$

So the new concentration of the diluted solution = old concentration × dilution factor.

Titrations

A titration is a method for finding the concentration of a solution by reacting a known volume with another solution of known concentration. Both the reacting volumes are measured as precisely as possible.

Using the balanced equation for the reaction, the unknown concentration of the solution can be found.

> 🖩 **Worked example: Finding the concentration of an alkali**
>
> In a titration, 25.0 cm³ of potassium hydroxide solution was pipetted into a conical flask. A 0.020 mol dm⁻³ solution of sulfuric acid was added from a burette. An indicator in the solution changed colour when 27.9 cm³ of sulfuric acid had been added. What is the concentration of the potassium hydroxide?
>
> | Formulae and balancing ratio | H_2SO_4 | 2KOH |
> |---|---|---|
> | volume / cm³ | 27.9 (0.0279 dm³) | 25.0 (0.0250 dm³) |
> | concentration / mol dm⁻³ | 0.020 | $\frac{1.116 \times 10^{-3}}{0.0250} = 0.045$ |
> | moles | 5.58×10^{-4} | 1.116×10^{-3} |
>
> **Step 1:** Write out the balanced equation for the reaction, in order to find the ratio of moles:
>
> $$H_2SO_4 + 2KOH \rightarrow K_2SO_4 + 2H_2O.$$
>
> **Step 2:** Calculate the number of moles of H_2SO_4:
>
> $$0.020 \times 0.0279 = 5.58 \times 10^{-4} \, mol$$
>
> **Step 3:** Use the balancing ratio to find the number of moles of KOH:
>
> $$2 \times 5.58 \times 10^{-4} = 1.116 \times 10^{-3} \, mol$$
>
> **Step 4:** Convert this number to a concentration using the volume of KOH:
>
> $$c = \frac{n}{v} = \frac{1.116 \times 10^{-3}}{0.0250} = 0.045 \, mol \, dm^{-3}$$

Summary questions

1 What is the concentration of these solutions of NaCl:
   a 0.200 mol in 250 cm³
   b 2.0×10^{-3} mol in 50 cm³
   c 2.5 mol in 20 dm³. *(3 marks)*

2 28.6 g of HCl was dissolved in water and made up to 200 cm³. 10 cm³ of this solution was then removed and diluted to 250 cm³ by adding distilled water:
   a What is the concentration in mol dm⁻³ of the original solution?
   b What is the concentration of the diluted solution? *(3 marks)*

3 20.0 cm³ of a solution of KOH was analysed in a titration against 0.0125 mol dm⁻³ sulfuric acid (H_2SO_4). An average titre of 16.73 cm³ of sulfuric acid was recorded. Calculate the concentration of the KOH in g dm⁻³. *(4 marks)*

Synoptic link

You will need to be able to write equations for the reactions of common acids and alkalis. Look at Topic 1.2, Balanced equations and Topic 8.1, Acids, bases, alkalis, and neutralisation, to help you.

Revision tip

In AS examinations you should find that the questions on titration calculations will be structured to help you with the steps in your working. In A-level examinations you will not be given this help.

Maths skill: Avoiding rounding errors

In multi-step calculations, like these titration calculations, it is important to avoid rounding your answers at an early stage. Keep the number in your calculator as you go through each stage of the calculation, and only round down to the appropriate number of significant figures at the end of the calculation.

Challenge

In some questions there may be extra steps in your calculation after you find the concentration of the reacting solution. For example, the solution used in the titration may have been diluted and you may need to use the dilution factor to find the original concentration of the undiluted solution.

The potassium hydroxide in the worked example was taken from an alkaline battery. 3.57 cm³ of the battery fluid was made up to 250 cm³ with distilled water for use in the titration. Calculate the concentration of potassium hydroxide in the original battery fluid.

Molar volumes of gases

One mole of any gas occupies the same volume as any other gas, as long as the conditions are the same.

This volume is called the molar volume, and for room temperature and pressure (around $20\,°C$ and 1 atmosphere pressure, or $10\,100\,Pa$), this is $24.0\,dm^3$.

Using molar volumes of gases

At RTP (room temperature and pressure), the amount in moles of a gas is given by:

$$\text{Moles} = \frac{\text{volume in } dm^3}{24.0}; \text{ so volume in } dm^3 = \text{moles} \times 24.0.$$

Calculating reacting masses and volumes

If the question involves only gases, you can use a simpler method. There is no need to convert the volumes into moles, because equal volumes of gases contain the same number of moles.

The ratio of the reacting volumes is the same as the ratio of moles.

Revision tip

The molar gas volume at RTP is provided on the data sheet.

Revision tip

If you are given a volume of a gas in cm^3, you will need to convert it to dm^3 to use this expression:

$1000\,cm^3 = 1\,dm^3$.

Revision tip

To convert between m^3 and dm^3 (or cm^3), you need to know that $1\,m^3 = 1000\,dm^3$.

So $1\,m^3 = 1\,000\,000\,cm^3$.

Revision tip

The value of R, the gas constant, is given in your data sheet.

Worked example: Calculating reacting volumes of gases

When propane is burnt in a good supply of air, it undergoes complete combustion:

$$C_3H_8(g) + 5O_2(g) \rightarrow 3CO_2(g) + 4H_2O(l)$$

4.5 dm³ of propane is burnt in air. Calculate the volume of oxygen needed.

Step 1: Find the ratio of moles oxygen: moles propane (5:1).

Step 2: Multiply the volume of propane by this ratio: $4.5 \times 5 = 22.5$ dm³.

The ideal gas equation

The connection between volume, pressure and temperature for a gas is given by the ideal gas equation:

$$pV = nRT$$

p = pressure in Pa V = volume in m³ n = number of moles

T = temperature in K R = the gas constant: $8.314\,J\,mol^{-1}\,K^{-1}$.

By rearranging the ideal gas equation, you can predict the pressure or volume of a gas, given all the other conditions.

Worked example: Calculating the volume of a gas at non-standard conditions

34.0 g of ammonia are heated to 373 K at a pressure of 150 000 Pa. Calculate the volume, in dm³, of this mass of ammonia.

Step 1: Rearrange the ideal gas equation to make V the subject:

$$V = \frac{nRT}{p}$$

Step 2: Calculate n, the number of moles of ammonia:

$$M_r(NH_3) = 17. \text{ So } n = \frac{34}{17} = 2.$$

Step 3: Substitute the values of n, R, T, and p into the equation:

$$V = \frac{2 \times 8.314 \times 373}{150\,000} = 0.0413 \text{ m}^3.$$

Step 4: Convert m³ to dm³:

$$1 \text{ m}^3 = 1000 \text{ dm}^3, \text{ so } V = 0.0413 \times 1000 = 41.3 \text{ dm}^3$$

Concentrations of gases

Percentage by volume

In mixtures of gases, such as the air in the atmosphere, the concentration of each gas is often given as the percentage by volume.

Parts per million

If a gas is present in a low concentration – less than 1% by volume – the concentration might be given in parts per million (ppm).

Maths skill: Rearranging equations

The ideal gas equation contains 5 different terms. You may be asked to rearrange it to make p, V, n, or T the subject of the equation. Practise rearranging the equation so that you are confident in rearranging in all these different ways.

Revision tip

To convert between % and ppm, remember that 100% = 1 000 000 ppm.
So 1% = 10 000 ppm.
To convert from % to ppm, multiply by 10 000.
To convert from ppm to %, divide by 10 000.

Summary questions

1 **a** The concentration by volume of neon in the atmosphere is 18.2 ppm. Express this as a %.

    b The carbon dioxide concentration is now greater than 0.04%. Convert this concentration into ppm.   *(2 marks)*

2 Hydrogen peroxide (H_2O_2) decomposes slowly to form oxygen gas:
$$2H_2O_2(aq) \rightarrow 2H_2O(l) + O_2(g)$$
What volume of oxygen gas is formed if 2.40 g of hydrogen peroxide decomposes at RTP?   *(3 marks)*

3 A sample of oxygen has a volume of 360 dm³ at a temperature of 400 K and 200 000 Pa. How many moles of oxygen molecules are present?   *(3 marks)*

Weighing a solid

Masses of solids are normally measured on an electronic balance that records masses to two or three decimal places.

Weighing out a known mass of solid

1 Place a weighing bottle or similar container on the balance.
2 Add the required mass of solid.
3 Reweigh the weighing bottle + solid.
4 Transfer the solid to the reaction vessel or volumetric flask.
5 Reweigh the weighing bottle without the solid.

Heating to constant mass

In some experiments you need to measure the mass of a solid remaining after thermal decomposition. Thermal decomposition of hydrated salts can be carried out in a crucible supported on a pipe-clay triangle.

Key term

Thermal decomposition: A reaction in which heat causes a compound to break down into simpler substances.

Synoptic link

You can see how to perform calculations using the data from thermal decomposition of hydrated salts in Topic 1.3, Using equations to work out reacting masses.

Weighing to constant mass ensures that decomposition is complete:

1 Weigh an empty crucible.
2 Add the hydrated salt and weigh the crucible + hydrated salt.
3 Heat the crucible strongly for several minutes.
4 Allow the crucible to cool and weigh the crucible and contents.

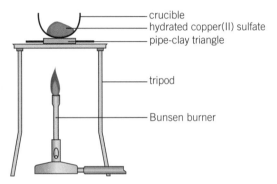

▲ **Figure 1** *Heating a hydrated salt in a crucible.*

5 Heat strongly for a further minute and reweigh the crucible and contents again.
6 If the mass has changed, then repeat this last step until two successive weighings are identical.

Volumetric equipment

Pipettes and burettes are described as volumetric equipment, because they can be used to measure volumes of liquids accurately.

Pipettes

Pipettes are used to precisely measure out fixed volumes of liquids (e.g. $10 \, cm^3$ or $25 \, cm^3$).

Liquid is sucked up into the pipette until it reaches the mark on the neck of the pipette.

The level of the liquid is viewed at eye level.

Burettes

Burettes are used to measure the volume of solution used to reach the end point of a reaction in a titration, or to measure out the volumes of solutions needed to produce a soluble salt.

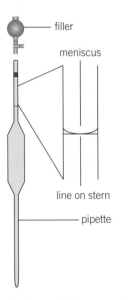

▲ **Figure 2** *Measuring the volume of liquid in a pipette*

Readings on burettes should be recorded to the nearest 0.05 cm³.

Titrations

This technique is used to determine the concentration of a solution by reacting it with another solution of known concentration.

Carrying out an acid/alkali titration

> ### Challenge
>
> Other types of reactions can be used as the basis of a titration. Sodium thiosulfate can be added to a solution of iodine using a starch indicator, or potassium manganate(VII) solution can be added to solutions of reducing agents, such as Fe^{2+} ions.
>
> The end point of a thiosulfate / iodine titration occurs when a blue-black colour is seen; the end point of a manganate titration occurs when a pink colour is seen. What substances are responsible for these colours?

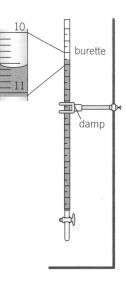

When the reaction is complete, it is signalled by the colour change of an indicator. The main stages are:

1   Pipette a known volume of the solution of unknown concentration into a clean conical flask.
2   Place the flask on a white tile – this improves the visibility of any colour changes.
3   Add a few drops of a suitable indicator – do not add too much, because the indicator reacts with acids and alkalis and so will interfere with your results.
4   Rinse a clean burette with the solution of known concentration.
5   Fill the burette with this solution.
6   Take an initial reading of the volume from the burette, using the lowest point of the meniscus of the liquid.
7   Swirl the conical flask and add the liquid from the burette 1 cm³ at a time until the expected colour change occurs. Keep swirling to mix the contents of the flask throughout the process.
8   Rinse the conical flask with distilled water and repeat steps 1, 2, 3, 5, and 6. This time, run in the solution from the burette up to 1 cm³ before the colour change obtained in step 7. Then, swirling thoroughly, add more solution dropwise until the desired colour change is observed. This gives an accurate result.
9   Repeat step 8 until you get three titre values within 0.1 cm³ of each other. These are called concordant results. The average of these titres is used in the calculations.

Preparing solutions

A standard solution can be made up either from a solid or from accurate dilution of another standard solution.

Making a standard solution from a solid

1   Weigh out the mass of solute required (see Weighing a solid, above).
2   Transfer the solid into a beaker and dissolve by stirring it with some distilled water.
3   Transfer this solution to a volumetric flask of a suitable volume.
4   Rinse the beaker and stirring rod with more distilled water and add these washings to the flask.

> ### Synoptic link
>
> Calculations using titration data are described in Topic 1.4, Concentrations of solutions.

> ### Revision tip
>
> To describe the colour change at the end point of a titration, you will need to know the colours of some common indicators. Phenolphthalein is magenta in alkali and colourless in acid. Methyl orange is red in acid and yellow in alkali.

> ### Key terms
>
> **Titre:** The volume of liquid delivered from a burette when the end-point of a reaction is reached.
>
> **Standard solution:** A solution whose concentration is accurately known.

Synoptic link

Calculating the concentration of a diluted solution is described in Topic 1.4, Concentrations of solutions.

5 Add distilled water to the flask to make it up to the mark. Use a dropping pipette for the last cm^3.

6 Invert the flask several times to ensure complete mixing.

Making a standard solution by dilution

A standard solution is diluted down by taking a small volume of the original solution and diluting it to a greater volume using distilled water.

1 A volumetric pipette is used to measure out the required volume of the original solution.

2 This is transferred to a volumetric flask of a suitable volume.

3 The flask is made up to the mark as described above.

Measuring volumes of gases

The volume of a gas produced in a chemical reaction can be measured using either a gas syringe or an inverted burette.

Gas syringe

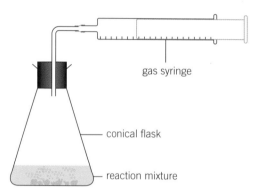

▲ **Figure 4** *Collecting a gas using a gas syringe*

● Up to $100\,cm^3$ of gas can be collected.

● The volume can be measured to the nearest $1\,cm^3$.

Using an inverted burette or measuring cylinder

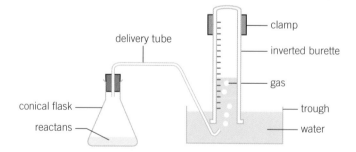

▲ **Figure 5** *Collecting a gas using a measuring cylinder*

● A burette can collect up to $50\,cm^3$ of gas; measuring cylinders can collect much greater volumes.

● This method is not suitable if the gas is soluble in water.

Chapter 1 Practice questions

1 What mass of sodium hydroxide, NaOH is needed to form $200\,cm^3$ of a $0.25\,mol\,dm^{-3}$ solution?

   A $2000\,g$

   B $2\,g$

   C $32\,g$

   D $10\,g$

   (1 mark)

2 Choose the phrase that correctly completes the following definition:

   The relative isotopic mass is the mass of one atom of an isotope compared to….

   A The mass of a carbon-12 atom.

   B $12\,g$ of carbon-12 atoms.

   C one-twelfth of $12\,g$ of carbon-12 atoms.

   D one-twelfth of the mass of a carbon-12 atom.

   (1 mark)

3 $1.40\,g$ of ethene, C_2H_4, reacts with excess steam, H_2O, to form ethanol, C_2H_5OH. The equation for the reaction is:

   $C_2H_4 + H_2O \rightarrow C_2H_5OH$

   $1.06\,g$ of ethanol are formed in the reaction. The % yield of this process is given by the equation:

   A $\dfrac{1.06}{1.40} \times 100$

   B $\dfrac{1.06}{\dfrac{(1.40 \times 28)}{46}} \times 100$

   C $\dfrac{1.06}{\dfrac{(1.40 \times 46)}{28}} \times 100$

   D $\dfrac{1.40}{\dfrac{(1.06 \times 28)}{46}} \times 100$

   (1 mark)

4 A compound that contains C, H, and O atoms only has the following % composition by mass: C: 40% O: 53%.

   The relative molecular mass of the compound was found to be 60.

   The molecular formula of the compound is:

   A CH_2O

   B CO

   C $C_2H_4O_2$

   D C_2O_2

   (1 mark)

5 Barium chloride solution was added to a solution of sodium sulfate. A white precipitate of barium sulfate was formed. The following are possible equations to describe what has happened. Which of these are correct?

   1 $BaCl(aq) + NaSO_4(aq) \rightarrow PrCl(aq) + BaSO_4(s)$

   2 $Na^+(aq) + Cl^-(aq) \rightarrow NaCl(s)$

   3 $Ba^{2+}(aq) + SO_4^{2-}(aq) \rightarrow BaSO_4(s)$

A 1, 2, and 3
B Only 1 and 2
C Only 2 and 3
D Only 3

(1 mark)

6 Hydrochloric acid in a sample of toilet cleaner was analysed.
   25 cm³ of the toilet cleaner was dissolved in water to make a solution of
   250 cm³.
   10.0 cm³ samples from this **diluted solution** were extracted using a
   volumetric pipette and titrated against 0.500 mol dm⁻³ sodium hydroxide.
   The average titre of sodium hydroxide was 15.75 cm³.
   The equation for the reaction between sodium hydroxide and hydrochloric
   acid is:

   NaOH (aq) + HCl(aq) → NaCl(aq) + H₂O(l).

   a Calculate the average number of moles of sodium hydroxide
      used in the titration. *(1 mark)*

   b Write down the number of moles of hydrochloric acid in a
      10.0 cm³ sample. *(1 mark)*

   c Calculate the concentration of hydrochloric acid, in mol dm⁻³,
      in the diluted solution to 3 s.d. *(2 marks)*

   d What was the concentration, in g dm⁻³, of hydrochloric acid
      in the original, undiluted, solution of toilet cleaner? *(2 marks)*

   e A technician made up 200 cm³ of the standard 0.500 mol dm⁻³ sodium
      hydroxide solution from a stock solution of 2 mol dm⁻³ sodium hydroxide.
      Describe how the technician could carry out this procedure, giving
      details of the equipment that would be used to ensure that
      the solution was made up to a high degree of accuracy. *(4 marks)*

7 Magnesium carbonate is a common mineral found on Earth.
   When it is heated it forms carbon dioxide gas:

   $$MgCO_3(s) \rightarrow MgO(s) + CO_2(s)$$

   2.00 g of magnesium carbonate was heated. The carbon dioxide gas was
   collected and its volume was measured.

   a Draw a labelled diagram showing how this experiment
      could be carried out. *(3 marks)*

   b Calculate the volume of gas that is formed in this experiment,
      measured at room temperature and pressure. *(2 marks)*

   c Magnesium carbonate has also been detected on the surface of Mars.
      Scientists expect that any magnesium carbonate present on the surface of
      Mars will be decomposing slowly to form carbon dioxide, which will be
      released into the Martian atmosphere.

      Calculate the volume, in dm³, that would be occupied by 1 mole of
      carbon dioxide in the Martian atmosphere. Assume that the pressure
      in the Martian atmosphere is 600 Pa and that the temperature is 210 K.

      (3 marks)

2.1 A simple model of the atom

Specification reference: EL(g), EL(a)

Inside the atom

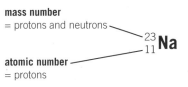

mass number
= protons and neutrons

atomic number
= protons

$^{23}_{11}$Na

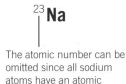

^{23}Na

The atomic number can be omitted since all sodium atoms have an atomic number of 11.

▲ **Figure 1** *A nuclear symbol*

The nuclear symbol tells you how many protons, electrons, and neutrons there are in an atom of an element:

- number of protons = atomic number (bottom number)
- number of electrons in a neutral atom = atomic number (bottom number)
- number of neutrons = mass number – atomic number (top number – bottom number).

So a sodium atom has 11 protons, 11 electrons, and 12 neutrons.

Atomic models

This model of an atom, known as the nuclear model, has developed over time. Models are tested using experimental investigations and are revised when observations are made that are not predicted by the model.

The succession of models

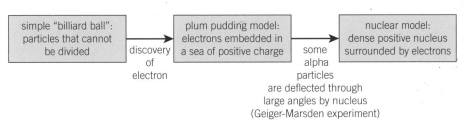

▲ **Figure 3** *Timeline for development of the atomic model*

Evidence from atomic spectra and the patterns of ionisation enthalpy led to a more sophisticated model of the atom known as the Bohr model.

The electrons are arranged in shells (also called energy levels).

---

protons and neutrons
in nucleus

electrons.

▲ **Figure 2** *The nuclear model of the atom*

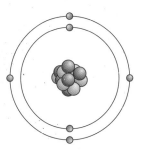

▲ **Figure 4** *The Bohr (electron shell) model of the atom*

Summary questions

1 The electron was discovered in 1897. Suggest one way in which this discovery caused the accepted model of the atom to change. (*1 mark*)

2 How many protons, neutrons and electrons are there in:
   a $^3_1$H   b $^{47}_{20}$Ca   c $^{23}_{11}$Na   (*3 marks*)

3 In the Geiger-Marsden experiment, alpha particles (helium nuclei) were fired at a thin sheet of gold atoms. Most of the alpha particles passed through the gold sheet but some of the alpha particles bounced back. Describe the features of the atomic model that were suggested by this observation. In your answer you should make it clear how the experimental evidence links with the model. (*2 marks*)

Nuclear fusion

Under certain conditions lighter nuclei can join together to form a single heavier nucleus.

High temperatures and/or pressures are required to provide the energy needed to overcome the repulsion between two positive nuclei.

The formation of new elements

Nuclear fusion is common in the centre of stars, where the necessary conditions of high temperature and pressure exist. In most stars, helium is being formed from hydrogen by nuclear fusion.

Heavier elements are formed in some particularly hot or heavy stars.

Nuclear equations

You need to be able to write nuclear equations for fusion reactions:

- In a nuclear equation, the total mass numbers and atomic numbers of the products and reactants must balance.
- Some fusion reactions produce one or more neutrons. The nuclear symbol for a neutron is $_0^1 n$; this should be included in the equation as a product.

> ### 🖩 Worked example: Writing nuclear equations
>
> Write a nuclear equation for the fusion of a $_1^1 H$ nucleus with a $_1^2 H$ nucleus to form a single heavier nucleus:
>
> **Step 1:** Write out the equation, leaving out the details of the product nucleus:
>
> $$_1^1 H + _1^2 H \longrightarrow {}_?^? ?$$
>
> **Step 2:** Add up the total mass number and atomic number on the left-hand side of the equation. These will be the mass number and atomic number of the product nucleus:
>
> $$_1^1 H + _1^2 H \longrightarrow {}_2^3 ?$$
>
> **Step 3:** Look up the symbol of the element with atomic number = 2 (He)
>
> $$_1^1 H + _1^2 H \longrightarrow {}_2^3 He$$

Summary questions

1 Nuclear fusion reactions require high temperatures and pressure. Why are these conditions necessary? *(2 marks)*

2 Complete these nuclear equations for fusion reactions by identifying particle X:

   a $_1^2 H + _1^3 H \longrightarrow {}_2^4 He + X$

   b $_6^{13} C + _2^4 He \longrightarrow X + {}_0^1 n$ *(2 marks)*

3 A fusion reaction occurs between two $_2^3 He$ to produce a different helium nucleus and two identical hydrogen nuclei. Write the nuclear equation for this reaction. *(1 mark)*

2.3 Shells, sub-shells, and orbitals

Specification reference: EL (e), EL (f)

- Electrons exist in shells and these are designated as $n = 1$, $n = 2$, $n = 3$, etc. The further away a shell is from the nucleus, the larger its n number.
- These shells are sub-divided into **sub-shells**, labelled s, p, d, and f.
- Notice that the $4s$ sub-shell is at a lower energy than the $3d$. This affects the order in which electrons fill up the sub-shells.

Orbitals

- Each sub-shell is further divided into atomic orbitals – each atomic orbital can hold a maximum of two electrons. These two electrons must have opposite (or paired) spins.
- s-orbitals and p-orbitals have different shapes.
- Orbitals are often represented by boxes. Drawing arrows in these boxes is a good way of representing the filling up of orbitals by electrons.

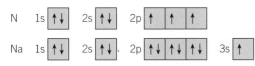

▲ **Figure 3** *A pair of electrons in an orbital. Electrons have a property known as 'spin' and the direction of the arrows show that the electrons have opposite spins*

Distribution of electrons in atomic orbitals

Shell	Description	Sub-shells
First	$n = 1$ has only one sub-shell	s
Second	$n = 2$ has two sub-shells	s and p
Third	$n = 3$ has three sub-shells	s, p, and d
Fourth	$n = 4$ has four sub-shells	s, p, d, and f

- An s sub-shell has 1 orbital holding a maximum of 2 electrons.
- A p sub-shell has 3 orbitals holding a maximum of 6 electrons.
- A d sub-shell has 5 orbitals holding a maximum of 10 electrons.

1 The orbitals are filled in order of increasing energy.

2 Where there is more than one orbital at the same energy, the orbitals are first occupied singly by electrons. When each orbital is singly occupied, then electrons pair up in orbitals.

3 Electrons in singly occupied orbitals have parallel spins.

4 Electrons in doubly occupied orbitals have opposite spins.

Representing electron distribution

The way that electrons are distributed between sub-shells is called the electron configuration of an atom or ion. You could be asked to write out an electronic configuration or to draw out the arrangement of electrons in boxes.

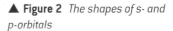

▲ **Figure 4** *The electronic configuration of N and Na atoms, showing electrons in boxes*

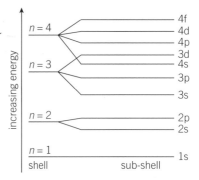

▲ **Figure 1** *Shells $n = 1$ to $n = 4$ and sub-shells*

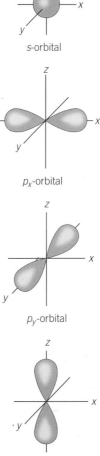

▲ **Figure 2** *The shapes of s- and p-orbitals*

⌨ Worked example: Electron configuration of a magnesium atom

Step 1: Sodium has an atomic number of 11, so there are 11 protons and 11 electrons in an Na atom.

Step 2: The first 2 electrons go into the lowest available sub-shell: $1s^2$.

Step 3: The next 2 electrons go into the next lowest sub-shell: $2s^2$.

Step 4: The next 6 electrons go into the next lowest sub-shell: $2p^6$.

Step 5: The total number of electrons added so far is 10; a Na atom has 11 electrons, so just one more needs to go into the next lowest sub-shell: $3s^1$.

Step 6: The electron configuration is $1s^2 2s^2 2p^6 3s^1$.

s, *p*, and *d* blocks

The periodic table is divided into blocks. The name of the block tells you the type of sub-shell occupied by the outermost (highest energy) electrons.

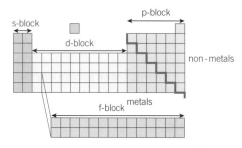

▲ **Figure 5** *The periodic table, showing the s, p, d, and f blocks*

⌨ Worked example: The outer sub-shell structure of arsenic (As)

Step 1: Arsenic is in the *p*-block. The outer sub-shell will be a *p* sub-shell.

Step 2: Arsenic is in the 3rd column of the *p*-block. The *p* sub-shell contains 3 electrons.

Step 3: Arsenic is in the 4th period. The outer sub-shell configuration is $4p^3$.

Summary questions

1   The orbitals in a $3p$ sub-shell can be represented as boxes, as shown in the following diagram. Copy and complete the diagram to show the arrangement of the $3p$ electrons in a sulfur atom. *(2 marks)*

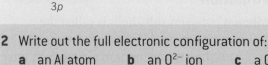
$3p$

2   Write out the full electronic configuration of:
    a   an Al atom   **b**   an O^{2-} ion   **c**   a Ca^{2+} ion. *(3 marks)*

3   **a**   Complete the electron configuration of a V atom: [Ar] ___ ___.
    b   Write out the electron configuration of the outer electron sub-shell of a Polonium (Po) atom. *(2 marks)*

1 The symbol for an ion of lithium is $_3^7\text{Li}^+$.

Which of the rows in the table correctly describes the numbers of the different sub-atomic particles in this ion? *(1 mark)*

	Protons	Neutrons	Electrons
A	3	4	2
B	3	4	3
C	4	3	3
D	4	3	2

2 Ernest Rutherford proposed a new model of the atom in 1911. In this model, a small dense nucleus is surrounded by a cloud of negative electrons. The evidence that best supports this model is:

   A The jumps observed in the pattern of successive ionisation enthalpies of atoms

   B The discovery of the electron

   C The deflection of alpha particles through large angles by gold nuclei

   D The lines seen in atomic emission spectra. *(1 mark)*

3 Which of the following equations represents a fusion reaction?

   1 $2^{12}_{6}\text{C}^{1}_{0}\text{n} \rightarrow {}^{23}_{12}\text{Mg}$ 2 $^{7}_{3}\text{Li}^{1}_{1}\text{H} \rightarrow 2^{4}_{2}\text{He}$,   3 $^{32}_{15}\text{P} \rightarrow {}^{0}_{-1}\text{e} + {}^{32}_{16}\text{S}$

   A 1, 2, and 3

   B Only 1 and 2

   C Only 2 and 3

   D Only 3 *(1 mark)*

4 Which of the following statements about *p*-orbitals is true?

   1 Each *p*-orbital can hold 6 electrons.

   2 *p*-orbitals have a spherical shape.

   3 Ground state electrons in helium atoms do not occupy *p*-orbitals.

   A 1, 2, and 3

   B Only 1 and 2

   C Only 2 and 3

   D Only 3 *(1 mark)*

5 Fusion reactions occur in centre of the Sun and other stars. High temperatures and pressures are necessary for fusion reactions.

   a **i** State what is meant by the term fusion reaction. *(2 marks)*

   ii Explain why high temperatures and pressures are necessary for fusion reactions to occur. *(2 marks)*

   b In the Sun, the most common fusion process involves isotopes of hydrogen. In other stars, helium nuclei are also involved in fusion processes.

   i Complete this equation for the fusion process that occurs in the Sun: _____ + _____ → $^{3}_{2}\text{He}$. *(2 marks)*

   ii Identify the nucleus X formed as a result of the fusion of helium nuclei: $3^{4}_{2}\text{He} \rightarrow \text{X}$. *(2 marks)*

   iii Suggest why this process occurs in some stars, but is not the main fusion process in the Sun. *(1 mark)*

6 This question is about the electron configurations of some atoms and ions.

   a The diagram below shows an incomplete diagram to show the arrangement of electrons in an atom of silicon (Si).

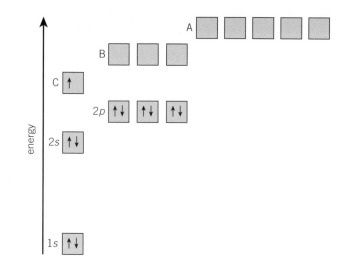

    i Complete the diagram to show the arrangement of
      electrons in an atom of silicon.   (*3 marks*)

    ii How many electrons can occupy the sub-shell labelled C
       on the diagram?   (*1 mark*)

    iii Identify the sub-shells labelled A and B in this diagram.
       A: _____ B: _____.   (*2 marks*)

    iv State one difference between the shape of an *s*-orbital
       and a *p*-orbital.   (*1 mark*)

b Complete the table to show the electron configurations of
  some atoms and ions.   (*4 marks*)

c Antimony (Sb) is in the *p*-block of the periodic table.

    i What does this information tell you about the
      arrangement of electrons in an antimony atom?   (*1 mark*)

    ii Write down the configuration of the outer sub-shell
       in an antimony atom.   (*1 mark*)

Species	Electron configuration
Na^+	. . . .
. . . .	$1s^2 2s^2 2p^5$
Cr	[Ar] . . . .

3.1 Chemical bonding

Specification reference: EL (i)

The types of bonding

The type of bond depends on the two atoms involved in the bond.

▼ Table 1

	Metal	Non-metal
Metal	metallic bonding	ionic bonding
Non-metal	ionic bonding	covalent bonding

Ionic bonding

The metal atom transfers electron(s) to the non-metal atom. This results in the formation of charged ions. The ions that are formed often have full outer shells, which makes them particularly stable.

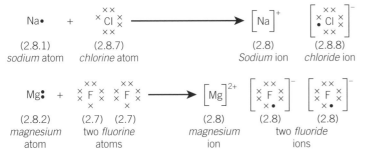

▲ **Figure 1** *Dot-and-cross diagrams for the formation of ionic compounds*

Ionic lattices and electrostatic attraction

- The cations (positive ions) and anions (negative ions) produced are held together in a giant ionic lattice.
- There is an electrostatic attraction between the cations and anions.

Covalent bonding

The two non-metal atoms involved in a covalent bond *share* one or more pairs of electrons. If two pairs of electrons are shared, then a double bond is formed.

Dative covalent bonds

Some molecules contain **dative covalent bonds**.

▲ **Figure 2** *Dative covalent bonds in molecules and ions*

Covalent bonds and electrostatic attraction

- There is an electrostatic attraction between the positive nuclei of the two atoms in the bond and the shared pair of negative electrons.
- This is greater than the repulsion between the two nuclei.

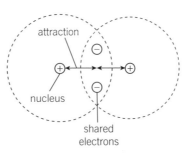

▲ **Figure 3** *The electrostatic attractions in a covalent bond*

▼ **Table 2** *Dot-and-cross structures of covalent molecules*

Molecular formula	Dot-and-cross diagram	Structural formula
H_2	H×H	H—H
NH_3	H×N with H above, H below	H—N with H above, H below
H_2O	H×O× with H	H—O with H
O_2	O×O	O=O
N_2	N×N×	N≡N
CO_2	O×C×O	O=C=O

Summary questions

1 What type of bonding is most likely between:
    a lithium and bromine,   **b** phosphorus and oxygen?   (*2 marks*)

2 Draw a *dot-and-cross* diagram to show the arrangement of outer shell electrons in the ionic compound magnesium oxide, MgO.   (*2 marks*)

3 In N_2O there is a central N atom bonded to a second N atom and an oxygen atom. Draw a *dot-and-cross* diagram to suggest a possible arrangement of electrons in this molecule.   (*2 marks*)

3.2 Shapes of molecules
Specification reference: EL (k)

Electron pair repulsion

- The shape of a molecule depends on the number of **groups** of electrons in the outer shell of the central atom.
- A group of electrons could be a bonding pair (a single bond), two bonding pairs (a double bond), three bonding pairs (a triple bond), or a lone pair.
- Groups of electrons repel each other.
- Groups of electrons will arrange themselves so as to be as far apart in space as possible, to minimise this repulsion.

The arrangement of electron pairs in molecules

You need to learn the 3-dimensional arrangements taken up by electron pairs in atoms with 2, 3, 4, 5, and 6 electron pairs in the outer shell. You will then be able to use the terms in this table to describe the shape of the molecule. You also need to know the bond angles that result from these arrangements.

Revision tip
Be careful how you describe electrons; you can describe the repulsion of electron pairs if there are no double or triple bonds in a molecule. If double or triple bonds are present, you should discuss the repulsion of electron groups.

Revision tip
The most common arrangement for electron pairs is tetrahedral, because many atoms in molecules have 8 electrons in their outer shell. You should learn any common exceptions to this, such as BF_3 and SF_6.

Number of pairs of electrons	2	3	4	5	6
Arrangement of electrons	Linear	Triangular planar	Tetrahedral	Trigonal bipyramid	Octahedral
Example	Beryllium chloride, $BeCl_2$	Boron trifluoride, BF_3	Methane, CH_4	Phosphorus pentachloride, PCl_5	Sulfur hexafluoride, SF_6
Diagram	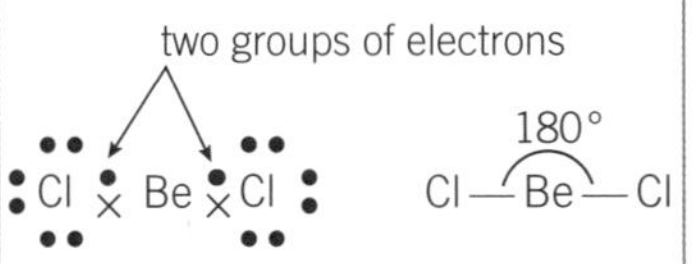	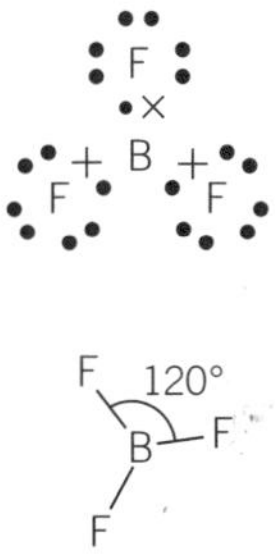	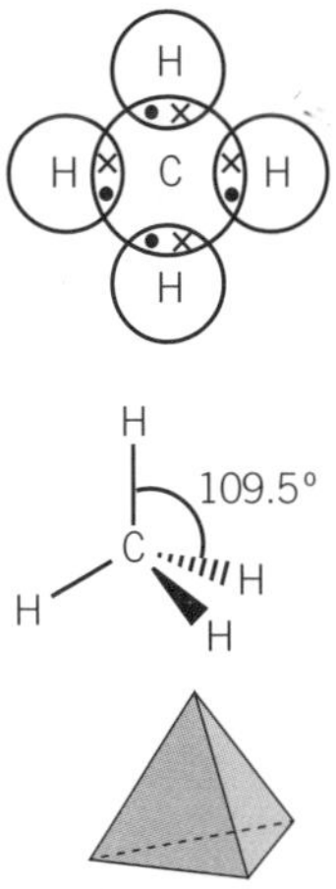	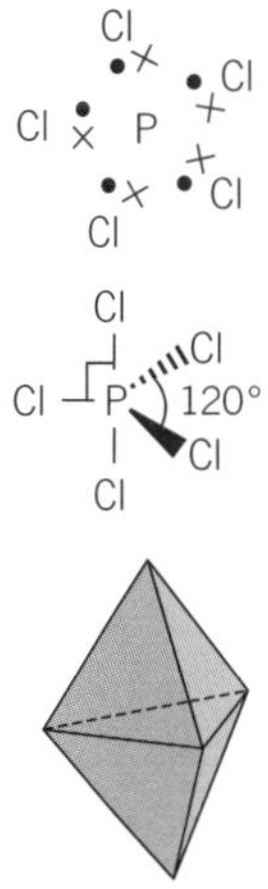	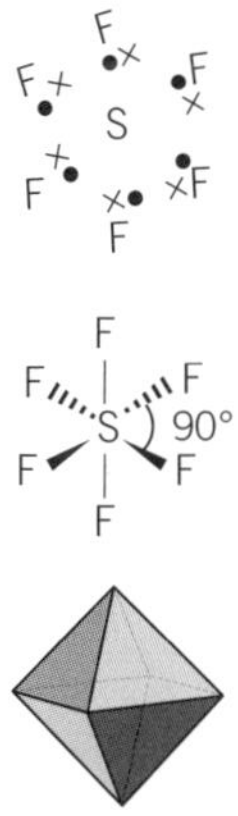
Bond angle	180°	120°	109.5	90 and 120°	90°

The effect of lone pairs
The shape of the molecule

If the molecule consists of a mixture of lone pairs and bonding pairs, you will need to use some different words to describe the shape of the molecule:

Key term

Lone pair: A pair of electrons in the outer shell of an atom that are not part of a covalent bond.

Pairs of electrons in outer shell	3 bonding pairs 1 lone pair	2 bonding pairs 2 lone pairs
Shape	Pyramidal	Bent (or V-shaped)
Example	Ammonia, NH_3	Water, H_2O
Diagram	ammonia	water

The bond angles

Lone pairs repel more strongly than bonding pairs of electrons. The effect of this is to force the bonding pairs slightly closer together, which makes the bond angle slightly smaller. On average, each lone pair present decreases the bond angle by about 2.5°.

Model answer: Use the idea of electron pair repulsion to describe the shape of a hydrogen sulfide molecule, H_2S. In your answer, you should predict and explain a value for the bond angle.

Electron pairs repel each other and get as far away as possible to minimise repulsion. The central S atom in the H_2S molecule has 4 electron pairs in its outer shell. These electron pairs take up a tetrahedral arrangement, but two of the electron pairs are lone pairs, so the shape of the molecule is V-shaped.

The angle between electron pairs in a tetrahedral arrangement is 109.5°, but the presence of two lone pairs decreases the angle between the bonding pairs by 5° (2 × 2.5°). So the bond angle is likely to be 104.5°.

Molecules with double and triple bonds

You can think of a double or triple bond as a single group of electrons. This often means that there are only 2 or 3 groups of electrons around a central atom.

Groups of electrons	2 double bonds	1 triple bond, 1 bonding pair	1 double bond, 2 bonding pairs
Shape	Linear	Linear	Triangular planar
Example	Carbon dioxide, CO_2	Ethyne, C_2H_2	Methanal, CH_2O
Diagram	$O=C=O$ 180°	$H-C≡C-H$ 180°	each 120°

Common misconception: Shape of molecules and arrangements of electrons

The name used to describe the arrangement of electron pairs is not always the same as the name given to the shape of the molecule.

The name given to the shape of the molecule refers to the arrangement of the atoms, not the electron pairs.

4 electron pairs will take up a tetrahedral arrangement; however, depending on the number of lone pairs present, the arrangement of the atoms can be tetrahedral, pyramidal, or bent.

Common misconception: Drawing organic molecules

Organic molecules are often drawn in a misleading way, showing 90° bond angles around some of the C atoms. Remember that 4 separate pairs of electron will arrange themselves tetrahedrally, with a bond angle of 109.5°.

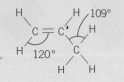

Synoptic link

You will need to be able to use the principles of electron pair repulsion to explain the shape and bond angles in organic molecules in Chapter 12, Organic chemistry: frameworks.

Summary questions

1 Name the 3-dimensional arrangement taken up by (a) 6 pairs of electrons (b) 3 pairs of electrons. Assuming that all the pairs are bonding pairs, suggest the bond angle in each case.   *(4 marks)*

2 Draw a *dot-and-cross* diagram for the molecule phosphorus trichloride, PCl_3. Use this diagram and the principle of electron pair repulsion to describe the shape of a PCl_3 molecule. In your answer you should predict and explain the bond angle in the molecule.   *(7 marks)*

3 a Suggest a *dot-and-cross* diagram for a sulfate ion, SO_4^{2-}, which contains an S atom surrounded by 4 O atoms, two of which have negative charges.
   b Draw a 3-dimensional diagram of the shape of a sulfate ion. In your answer you should predict and explain the bond angle in the molecule.   *(5 marks)*

Chapter 3 Practice questions

1   In which of these compounds is the bonding most likely to be ionic?

   A NO_2

   B MgH_2

   C HCl

   D BF_3 (*1 mark*)

2   Covalent bonds cause a force of attraction between atoms. The best
   explanation of this is:

   A The atoms now have a full outer shell and so are more stable.

   B There is electrostatic attraction between nuclei and the electron pair.

   C The two electrons in the bond attract one another.

   D There is electrostatic attraction between positive and negatively
            charged atoms.   (*1 mark*)

3   Which of these molecules is most likely to contain a dative covalent bond,
   in order to achieve noble gas structures for all the atoms in the molecule?

   A CO_2

   B CO

   C N_2

   D H_2CO (*1 mark*)

4   The *dot-and-cross* diagram of the molecule BF_3 is shown.

   Which of these statements about the molecule are true?

      1 It has a pyramidal structure.

      2 The bond angle will be 120°.

      3 The molecule will be planar.   (*1 mark*)

   A 1,2, and 3

   B Only 1 and 2

   C Only 2 and 3

   D Only 3   (*1 mark*)

5   What name is given to the shape of a molecule of sulfur hexafluoride,
   with 6 electron pairs in the outer shell of the sulfur atom?

   A Octahedral   **B** Tetrahedral

   C Hexahedral   **D** Hexagonal   (*1 mark*)

6   Ammonium chloride, NH_4Cl is an ionic compound used as a fertiliser. The two
   ions present in the compound are the ammonium ion, NH_4^+, and the chloride
   ion, Cl^-. At room temperature, they form a 3-dimensional ionic lattice.

   a **i** Draw out a diagram to show the arrangement of these ions in a
            layer of the ionic lattice.   (*2 marks*)

      ii Describe the forces that hold this lattice together at room
            temperature.   (*2 marks*)

   b The ammonium ion contains covalent bonds. The *dot-and-cross* diagram
            of the ammonium ion is shown.

      i Explain how this *dot-and-cross* diagram shows the presence of a
            dative covalent bond in the ammonium ion.   (*2 marks*)

      ii Use the *dot-and-cross* diagram to predict the shape and bond angle of
            the ammonium ion.   (*5 marks*)

   c In the soil, ammonium ions can be broken down into nitrogen gas,
            N_2, by certain types of bacteria.

      Draw a *dot-and-cross* diagram to show the arrangement of electrons in a
            molecule of nitrogen gas.   (*2 marks*)

Exothermic: A reaction that gives out energy and heats the surroundings.

Endothermic: A reaction that takes in energy and cools the surroundings.

Enthalpy changes

The enthalpy change, $\Delta_r H$, for a reaction is the quantity of energy transferred to or from the surroundings when the reaction is carried out in an open container.

Exothermic and endothermic reactions

- An exothermic reaction gives out energy from the system to the surroundings. The temperature of the surroundings increases, $\Delta_r H$ is negative.
- An endothermic reaction takes energy into the system from the surroundings. The temperature of the surroundings decreases, $\Delta_r H$ is positive.

Bond breaking and bond forming

- Breaking chemical bonds requires energy and so is an endothermic process.
- Forming chemical bonds releases energy and so is an exothermic process.

The overall effect of bond breaking and forming

Many reactions involve both bond breaking:

- If the energy released by forming bonds is greater than the energy required to break bonds, then the reaction will be **exothermic.**
- If the energy released by forming bonds is less than the energy required to break bonds, then the reaction will be **endothermic.**

Standard enthalpy changes

Several types of enthalpy change are particularly important. These are given specific names and symbols. If an enthalpy change is measured under standard conditions, it is known as a standard enthalpy change:

- (standard) enthalpy change of reaction ($\Delta_r H$), $2NaOH(aq) + H_2SO_4(aq) \rightarrow Na_2SO_4(aq) + 2H_2O(l)$; $\Delta_r H^\ominus = -115\,kJ\,mol^{-1}$
- $\Delta_c H$ (standard) enthalpy change of combustion ($\Delta_c H$), $CH_4(g) + 2O_2(g) \rightarrow CO_2(g) + 2H_2O(l)$; $\Delta_c H = -890\,kJ\,mol^{-1}$
- (standard) enthalpy change of formation ($\Delta_f H$), $H_2(g) + \frac{1}{2}O_2(g) \rightarrow H_2O(l)$; $\Delta_f H = -286\,kJ\,mol^{-1}$
- (standard) enthalpy change of neutralisation ($\Delta_{neut} H$), $H+(aq) + OH^-(aq) \rightarrow H_2O(l)$; $\Delta_{neut} H = -58\,kJ\,mol^{-1}$

Enthalpy changes and moles

You can calculate the amount of energy released or taken in during a reaction by using values for enthalpy changes of reaction, $\Delta_r H$, and data about the number of moles of the reactants.

Synoptic link

You may need to use data about the strength of covalent bonds to predict the $\Delta_r H$ value for a reaction. See Topic 4.3, Bond enthalpies, for more guidance.

> 🖩 **Worked example: Enthalpy change for an acid-base reaction**
>
> Sodium hydroxide solution reacts with sulfuric acid in an exothermic reaction:
>
> $2NaOH(aq) + H_2SO_4(aq) \rightarrow Na_2SO_4(aq) + 2H_2O(l)$ $\Delta_r H^\ominus = -115\,kJ\,mol^{-1}$
> Calculate the energy released when 100 cm³ of 2.00 mol dm⁻³ NaOH reacts with excess H_2SO_4.
>
> **Step 1:** Calculate the number of moles of NaOH $= \dfrac{100}{1000} \times 2.00 = 0.200\,mol$.
>
> **Step 2:** The $\Delta_r H^\ominus$ value gives the enthalpy change for the number of moles of the equation: number of moles of NaOH in the equation = 2.

Step 3: Find the ratio between the number of moles that actually react and number of moles in the equation: $\frac{0.2}{2} = 0.1$.

Step 4: Multiply the $\Delta_r H$ by this ratio: $-115 \times 0.1 = -11.5 \, \text{kJ}$.

11.5 kJ of energy will be released in this reaction (released because the sign is negative).

Calculating enthalpy changes from experimental results

Experiments to measure enthalpy changes usually involve transferring energy to or from water.

The amount of energy transferred to or from a mass of water is given by the equation:

E, energy transferred = m (mass of water) × c (specific heat capacity of water) × ΔT (temperature change of the water).

🖩 Worked example: Calculating enthalpy changes of combustion

0.880 g of heptane (C_7H_{16}) undergo complete combustion and the energy released is used to heat 250 cm³ of water. The temperature of the water rises by 19.0 °C. Calculate the $\Delta_c H$ of heptane in kJ mol⁻¹. Give your answer to 3 significant figures.

Step 1: Use $E = m \, c \, \Delta T$ to calculate the energy transferred to the water:
$E = 250 \times 4.18 \times 19 = 19\,855 \, \text{J}$.

(250 cm³ water is assumed to have a mass of 250 g, because the density of water is 1.00 g cm⁻³.)

Step 2: Convert the energy released into kJ $= \frac{19855}{1000} = 19.855 \, \text{kJ}$.

Step 3: Calculate the number of moles of heptane burnt: moles $= \frac{\text{mass}}{M_r} = \frac{0.880}{100}$ $= 0.0880 \, \text{mol}$.

Step 4: Scale the energy released to find the energy that would be released if 1 mole was burnt: $= \frac{\text{energy released in experiment}}{\text{number of moles burnt}} = \frac{19.855}{0.0880} = 2\,256.25 \, \text{kJ mol}^{-1}$.

Step 5: Write down the $\Delta_c H$ value with the correct sign, units and significant figures: $\Delta_c H = -2\,260 \, \text{kJ mol}^{-1}$.

🖩 Worked example: Calculating enthalpy changes in solution

An excess of zinc is added to 100 cm³ of 1.00 mol dm⁻³ copper(II) sulfate. The maximum temperature rise measured was 36.0 °C.

Calculate $\Delta_r H$ for this reaction in kJ per mole of copper sulfate. Give your answer to 3 s.f.

Step 1: Use $E = m \, c \, \Delta T$ to calculate the energy transferred to the water:
$E = 100 \times 4.18 \times 36.0 = 15\,048 \, \text{J}$
(100 cm³ solution is assumed to contain 100 g water).

Step 2: Convert the energy released into kJ: $= \frac{15048}{1000} = 15.048 \, \text{kJ}$.

Step 3: Calculate the number of moles of copper sulfate reacted: moles $= \frac{100}{1000} \times 1 = 0.1 \, \text{mol}$.

Step 4: Scale the energy released to find the energy that would be released if one mole was burnt: $= \frac{\text{energy released in experiment}}{\text{number of moles reacted}} = \frac{15.048}{0.1} = 150.48 \, \text{kJ mol}^{-1}$.

Step 5: Write down the $\Delta_r H$ value with the correct sign, units and significant figures: $\Delta_r H = -150 \, \text{kJ mol}^{-1}$.

Revision tip

Make sure that you are careful about the wording when you are comparing bond breaking and bond forming. One process requires energy and the other releases energy. So answers such as 'bond breaking requires more energy than bond forming' are not correct.

Key terms

Standard conditions: 1 atmosphere pressure (101 kPa), 298 K (25°C) and a concentration of 1 mol dm⁻³ for any solutions.

Standard enthalpy change of reaction ($\Delta_r H$): The enthalpy change for a reaction (described by an equation) that occurs between the number of moles of the reactants specified in the equation.

Standard enthalpy change of combustion $\Delta_c H$: The enthalpy change when 1 mole of a substance burns completely in oxygen under standard conditions.

Standard enthalpy change of formation, $\Delta_f H$: The enthalpy change when 1 mole of a substance is formed from its constituent elements in their standard states under standard conditions.

(Standard) enthalpy change of neutralisation ($\Delta_{neut} H$): The enthalpy change when 1 mole of hydrogen ions react with one mole of hydroxide ions to form 1 mole of water under standard conditions and in solutions with a concentration of 1 mol dm⁻³.

Synoptic link

You need to be able to describe how to obtain results to find energy transferred in chemical reactions. You can find this information in Experimental techniques.

Revision tip

The value of c, the specific heat capacity of water, is $4.18\ J\ g^{-1}\ K^{-1}$.

Challenge

In some experiments, energy is transferred to a solution of an ionic compound, such as copper sulfate solution. The volume of the solution is measured. What assumptions are necessary to complete the calculation for the amount of energy transferred? Do you think these assumptions are justified?

Summary questions

1 All combustion reactions are exothermic.

a State what you can conclude about:

i the sign of $\Delta_r H$ for a combustion reaction

ii the temperature change that will occur in the surroundings when combustion occurs. *(2 marks)*

b Explain why combustion reactions are exothermic, using ideas about bonding in your answer. *(2 marks)*

2 1.60 g of methanol (CH_3OH) is burnt in a spirit burner and used to heat 150 cm^3 of water. The maximum temperature rise recorded was 41.0 °C. Calculate a value for $\Delta_c H$ of methanol, in $kJ\ mol^{-1}$. Give your answer to 3 s.f. *(4 marks)*

3 Nitrogen triiodide, NI_3, decomposes in an exothermic reaction:

$2NI_3(s) \rightarrow N_2(g) + 3I_2(g)\ \Delta_r H^\ominus = -290\ kJ\ mol^{-1}$.

Calculate the energy released by this reaction if 1 kg of NI_3 decomposes. *(3 marks)*

4.2 Enthalpy cycles

Specification reference: DF (g)

Hess' Law

Hess' Law states that as long as the starting and finishing points are the same, the enthalpy change for a chemical reaction will always be the same, no matter how you go from start to finish. It is useful for calculating unknown enthalpy changes from ones for which data is available.

Hess' Law calculations from $\Delta_f H$ data

The key to Hess' Law calculations is to construct a triangular enthalpy cycle, to show the relationship between the various enthalpy changes. If $\Delta_f H$ data is given, then the substances at the bottom of the triangle will be elements in their standard states.

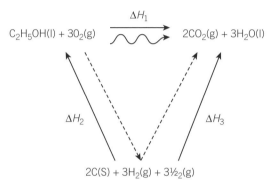

▲ **Figure 1** *Enthalpy cycle for Hess' Law calculation using $\Delta_f H$ data*

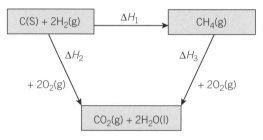

▲ **Figure 2** *Enthalpy cycle for Hess' Law calculation using $\Delta_c H$ data*

> **Revision tip**
> Remember that if an equation is reversed, the sign for ΔH is also reversed.

> **Revision tip**
> Remember to multiply $\Delta_f H$ by the number of moles in the equation.

> **Revision tip**
> You can calculate an enthalpy change for a reaction without drawing an enthalpy cycle by remembering that $\Delta_r H = \Delta_f H(\text{products}) - \Delta_f H (\text{reactants})$.

🖩 Worked example: Hess' Law calculation using $\Delta_f H$ data

Use the data provided to calculate a value for the enthalpy change represented by the equation:

$$C_2H_5OH(l) + 3O_2(g) \rightarrow 2CO_2(g) + 3H_2O(l)$$

Compound	$\Delta_f H$(kJ mol^{-1})
$C_2H_5OH(l)$	−277
$CO_2(g)$	−394
$H_2O(l)$	−286

Step 1: Construct an enthalpy cycle showing the reactants and products of the reaction as well as elements in their standard states. Make sure that the arrows point in the correct directions to represent the enthalpy changes.

Step 2: Choose suitable labels to represent each arrow (ΔH_1, ΔH_2, etc.).

Step 3: Calculate ΔH_2: $\Delta H_2 = \Delta_f H (C_2H_5OH(l)) = -277$ kJ mol^{-1}.

> **Common misconception: Missing data for $\Delta_f H$?**
> Students often worry that they cannot find data to complete Hess' Law calculations. But, of course, the enthalpy change of formation of an element in its standard state must be zero, by definition, because forming an element from the same element involves no chemical change. So $\Delta_f H (O_2)$ can be left out of the calculation.

◀

Revision tip

Make sure that the arrows in your enthalpy cycle point in the correct direction. If $\Delta_f H$ data is given, elements will appear at the bottom of the cycle and the arrows will point up; if $\Delta_c H$ data is given, combustion products will appear at the bottom of the cycle and arrows will point down.

Maths skill: Checking the sign

Think carefully about your answer, to check whether it is mathematically sensible. If the reaction you are dealing with is a combustion reaction, this will be exothermic. Your answer should have a negative sign. If the sign does not seem sensible, then check your working, particularly where you are handling numbers with negative signs. It is easy to accidently leave some of these signs out of your calculation.

Revision tip

Always include a sign in any value of ΔH that you give in an answer even if the value is positive.

Step 4: Calculate ΔH_3: $\Delta H_3 = 2 \times \Delta_f H\ (CO_2(g)) + 3 \times \Delta_f H\ (H_2O(l)) = (2 \times -394) + (3 \times -286) = -1646\ k\ mol^{-1}$.

Step 5: Use Hess' Law to write down a relationship between $(\Delta H_1, \Delta H_2,$ and $\Delta H_3)$: $\Delta H_1 = \Delta H_3 - \Delta H_2$.

Step 6: Plug the calculated values for ΔH_2 and ΔH_3 into the relationship to find ΔH_1: $\Delta H_1 = -1646 - (-277) = -1369\ kJ\ mol^{-1}$.

🖩 Worked example: Hess' Law calculation using $\Delta_c H$ data

Calculate the enthalpy change for the reaction $C(s) + 2H_2(g) \rightarrow CH_4(g)$.

Substance	$\Delta_c H$ / kJ mol^{-1}
$C(s)$	−394
$H_2(g)$	−286
$CH_4(g)$	−890

Step 1: Construct an enthalpy cycle showing the reactants and products of the reaction as well as combustion products (CO_2 and H_2O). Make sure that the arrows point in the correct directions to represent the enthalpy changes.

Step 2: Add in the correct number of oxygen molecules to make the combustion equations balance, as shown in the cycle above.

Step 3: Choose suitable labels to represent each arrow ($\Delta H_1, \Delta H_2,$ etc.).

Step 4: Calculate ΔH_2: $\Delta H_2 = \Delta_c H(C(s)) + 2 \times \Delta_c H(H_2) = -394 + (2 \times -286) = -966\ kJ\ mol^{-1}$.

Step 5: Calculate ΔH_3: $\Delta H_3 = \Delta_c H\ (CH_4(g)) = -890\ kJ\ mol^{-1}$.

Step 6: Use Hess' Law to write down a relationship between $(\Delta H_1, \Delta H_2,$ and $\Delta H_3)$: $\Delta H_1 = \Delta H_2 - \Delta H_3$.

Step 7: Plug the calculated values for ΔH_2 and ΔH_3 into the relationship to find ΔH_1: $\Delta H_1 = -966 - (-890) = -76\ kJ\ mol^{-1}$.

Summary questions

1 What $\Delta_f H$ data would you need in order to calculate the enthalpy change for the reaction: $C_2H_4(g) + H_2(g) \rightarrow C_2H_6(g)$? (*2 marks*)

2 Hydrazine, N_2H_4 reacts with oxygen to form N_2 and H_2O: $N_2H_4(l) + O_2(g) \rightarrow N_2(g) + 2H_2O(l)$.
   Use the following $\Delta_f H$ values to calculate a value for the $\Delta_r H$ of the reaction of hydrazine with oxygen:
   $\Delta_f H$ values / kJ mol^{-1}: $N_2H_4(l)$ + 50.6, $H_2O(l)$ −285.8. (*4 marks*)

3 The enthalpy change of formation of propane, $C_3H_8(g)$ can be found using $\Delta_c H$ data for propane and the elements carbon and hydrogen.
   a Draw out a fully labelled Hess's Law cycle that would enable you to calculate the $\Delta_f H$ of propane by this method.
   b Use the following data to calculate a value for the $\Delta_f H$ of propane:
   $\Delta_c H\ [C(s)] = -394\ kJ\ mol^{-1}$, $\Delta_c H[H_2(g)] = -286\ kJ\ mol^{-1}$
   $\Delta_c H\ [C3H_8(g)] = -2\ 219\ kJ\ mol^{-1}$. (*6 marks*)

4.3 Bond enthalpies

Specification reference: DF (e), DF (g)

Bond enthalpies

Bond enthalpy is a measure of the strength of a covalent bond; the greater the bond enthalpy, the stronger the bond. Short bonds are stronger than long bonds, so they have a greater bond enthalpy.

Bond enthalpies and enthalpy changes of reaction

You can use bond enthalpy data to calculate a value for the enthalpy change of a reaction involving covalent molecules.

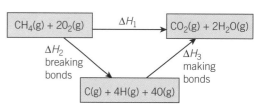

▲ **Figure 1** *Using Hess' Law to find the enthalpy change of a reaction:* $\Delta H_1 = \Delta H_2 + \Delta H_3$

🖩 Worked example: Enthalpy change from bond enthalpy data

Calculate the enthalpy change of combustion for ethane (C_2H_6) using the bond enthalpy data from the table in Table 1:

$$C_2H_6 + 3\tfrac{1}{2}O_2 \rightarrow 2CO_2 + 3H_2O$$

Step 1: Draw out diagrams to show the structures of the reactants and products.

Step 2: List the bonds broken and formed:

Bonds broken	Bonds formed
$1 \times C-C$ (347) = +347	
$6 \times C-H$ (413) = +2478	$4 \times C=O$ (805) = 3220
$3.5 \times O=O$ (498) = +1743	$6 \times O-H$ (464) = 2784
Total = **+4568**	Total = **−6004**
(+ sign because bond breaking)	(− sign because bond making)

Step 3: Multiply the number of bonds broken and formed by the bond enthalpies to calculate the total enthalpy change for bond breaking and bond forming.

Step 4: Add together the enthalpy changes for bond breaking and bond forming:

$\Delta H = +4568 - 6004 = -1436 \, kJ \, mol^{-1}$.

Problems with this calculation

The value obtained by this calculation is not the same as the data book value for $\Delta_r H$. There are two main reasons for this:

1. Average bond enthalpies are used in the calculation; the actual bond enthalpies in particular molecules will be slightly different.

2. Bond enthalpy data are for gaseous molecules; some molecules may be in a liquid state at 298 K.

Common misconception: Adding up the bonds

There are a number of common errors to beware of:

- Miscounting the number of C−C bonds in a molecule: C_2H_6 has 1 C−C bond, not 2.

- In a combustion reaction, forgetting to include the bonds in the oxygen molecules.

- Treating the bonds in CO_2 as C−O, not C=O.

Key term

Bond enthalpy: The energy required to break 1 mole of a particular bond, averaged over a range of different gaseous compounds.

▼ **Table 1**

Bond	Bond enthalpy / kJ mol⁻¹
C−C	+347
C−H	+413
O=O	+498
O−H	+464
C=O (in CO_2)	+805
C−O	+358
H−H	+436
C=C	+612

Summary questions

1. Describe how C−C and C=C bonds differ in:
   a. bond strength, b. bond length.   *(2 marks)*

2. a. Calculate a value for the $\Delta_c H$ of methane, CH_4 using bond enthalpy data.
   b. Suggest why this value differs from the data book value for $\Delta_c H$ of methane.   *(4 marks)*

3. Cyclopropane, C_3H_6, has a structure in which the three carbon atoms are arranged in the form of a ring. The $\Delta_c H^{\ominus}$ of cyclopropane is −2091 kJ mol⁻¹. Use this value and other bond enthalpy data from this page to deduce a value for the bond enthalpy of C−C in cyclopropane. Comment on your answer.   *(7 marks)*

Synoptic link

This Hess's cycle can also be drawn as a reaction profile (enthalpy level diagram), which will help you to understand the idea of activation energy. See Topic 10.2, The effect of temperature on rate.

Measuring energy transferred in a reaction

You can carry out experiments in which energy is transferred to or from a known mass of water.

Determining enthalpy changes of combustion of flammable liquids

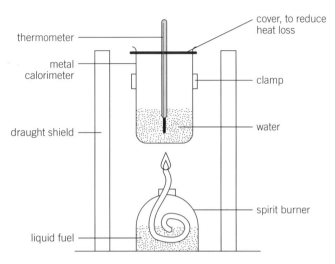

▲ **Figure 1** *Simple apparatus for measuring $\Delta_c H$*

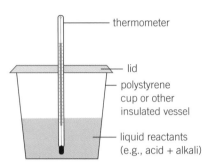

▲ **Figure 2** *Apparatus to determine ΔrH for a reaction in solution*

Record the initial temperature of the water and the maximum temperature reached when a known volume of water is heated by the complete combustion of a measured mass of fuel. The energy transferred = $m \times c \times \Delta T$. You can now calculate the enthalpy change for the combustion of 1 mole of the fuel used.

Determining enthalpy changes for reactions in solution

This method can be used for reactions between two solutions or between a solid and a solution. The energy is transferred to or from the water in the solution. The reaction is carried out in an insulated container, such as a polystyrene cup, fitted with a lid.

Record the volumes and initial temperature of the solution(s) and the maximum or minimum temperature reached during the reaction. The energy transferred = $m \times c \times \Delta T$. You can now calculate the enthalpy change for the reaction of the number of moles of reactants specified in the equation.

Using cooling curves for reactions involving solutions

You can adapt the experiment to correct for heat loss by taking measurements of the temperature over several minutes and plotting a cooling curve. This produces a more accurate value for ΔT.

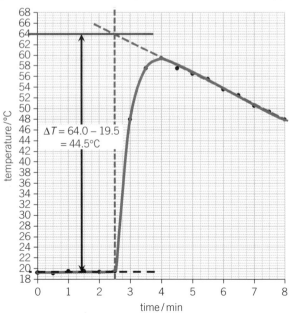

▲ **Figure 3** *Using a cooling curve to calculate a corrected value for ΔT (reaction started after 2½ minutes)*

Chapter 4 Practice questions

1 For which of these processes is the enthalpy change certain to be endothermic?

(handwritten: making exothermic breaking end)

  A $Cl_2(g) \rightarrow 2Cl(g)$

  B $Cl_2(g) + CH_4(g) \rightarrow CH_3Cl(g) + HCl(g)$

  C $NH_3(g) + BF_3(g) \rightarrow NH_3BF_3(g)$

  D $H_2(g) + Cl_2(g) \rightarrow 2HCl(g)$ *(1 mark)*

2 The combustion of methane is exothermic. The best explanation of this is:

  A The reaction involves only bond formation.

  B The energy released by bond forming is greater than the energy released by bond breaking. ✓

  C The energy released by bond forming is greater than the energy required for bond breaking.

  D The energy required to form bonds is less than the energy released by breaking bonds.   *(1 mark)*

(handwritten: Bond breaking – bond forme = –   □   –   □)

3 Which of these processes represents the reaction that occurs during the measurement of the enthalpy change of formation of $AgCl(s)$?

  A $Ag^+ (aq) + Cl^-(aq) \rightarrow AgCl(s)$

  B $Ag(s) + \frac{1}{2}Cl_2(g) \rightarrow AgCl(s)$

  C $2Ag(s) + Cl_2(g) \rightarrow 2AgCl(s)$

  D $Ag(g) + Cl(g) \rightarrow AgCl(s)$ *(1 mark)*

4 A known mass of methanol is burned in a plentiful supply of oxygen. The energy released is transferred to water and the results are used to calculate the $\Delta_c H$ of methanol. The value obtained is much less negative than the databook value for $\Delta_c H^\ominus$. The most likely explanation of this is:

  A Heat transferred from the surroundings to the water.

  B Incomplete combustion of the fuel.

  C Heat transferred from the water to the surroundings.

  D Carrying out the reaction under non-standard conditions.   *(1 mark)*

5 The Hess's cycle can be used to determine the value of ΔH_1.

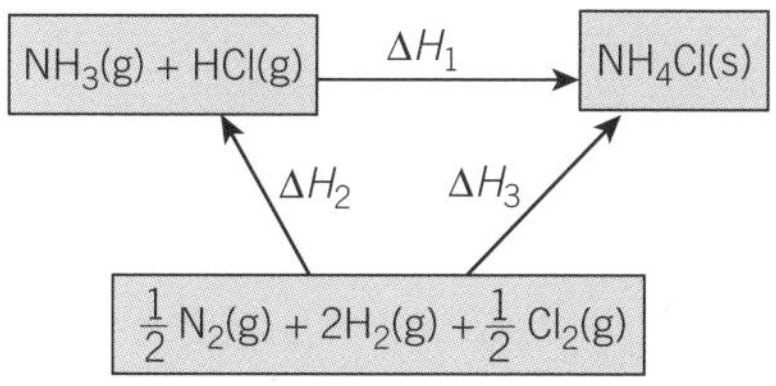

  The value of ΔH_1 is given by the expression:

  A $\Delta_f H\ [NH_4Cl] - \Delta_f H\ [NH_3] + \Delta_f H\ [HCl]$

  B $\Delta_f H\ [NH_4Cl] - \Delta_f H\ [NH_3] - \Delta_f H\ [HCl]$

  C $\Delta_f H\ [NH_3] + \Delta_f H\ [HCl] - \Delta_f H\ [NH_4Cl]$

  D $\Delta_f H\ [NH_4Cl] - \frac{1}{2}\Delta_f H\ [N_2] - 2\,\Delta_f H\ [H_2] - \frac{1}{2}\Delta_f H\ [Cl_2]$ *(1 mark)*

(diagram: box $NH_3(g) + HCl(g)$ → ΔH_1 → box $NH_4Cl(s)$; arrows ΔH_2 and ΔH_3 down/up to box $\frac{1}{2} N_2(g) + 2H_2(g) + \frac{1}{2} Cl_2(g)$)

(handwritten: $\Delta H_1 = -\Delta H_2 + \Delta H_3$)

6 Ethanol, C_2H_5OH, can be used as a liquid fuel. The enthalpy of combustion of ethanol can be determined by burning it in a spirit burner and using the energy released to heat water.

  The following results were obtained in an experiment:

(handwritten: $\Delta H_1 = (-NH_3) + (HCl) + NH_4Cl$)

Mass of calorimeter	120 g
Mass of calorimeter + water	220 g
Mass of spirit burner at start	43.56 g
Mass of spirit burner at end	42.46 g
Initial temperature of water	20°C
Maximum temperature of water	44°C

(handwritten: $Q = mc\Delta T$ $Q =$)

  a Calculate the energy, in J, transferred to the water in the calorimeter.   *(2 marks)*

b i Use your answer to **a** to determine a value for the enthalpy of combustion of ethanol in kJ mol⁻¹. Give your answer to 3 s.f. (*4 marks*)

ii The databook value for the standard enthalpy of combustion of ethanol is much more negative than values determined by experiments. Suggest two reasons for this. (*2 marks*)

7 The structure of butanone is shown below:

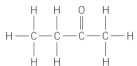

Butanone can undergo complete combustion in a plentiful supply of air:

$$C_4H_8O(l) + 6O_2(g) \rightarrow 4CO_2(g) + 4H_2O(l)$$

a i Use the bond enthalpy data below to calculate a value for the enthalpy change of combustion of butanone.

Bond	Average bond enthalpy / kJ mol⁻¹
C–C	+347
C–H	+413
C=O	+805
O–H	+464
O=O	+498

(*4 marks*)

ii The databook value for the enthalpy change of combustion of butanone is significantly different to the value that is calculated using these bond enthalpies. State two reasons for this difference (*2 marks*)

b Butanal also has the molecular formula C_4H_8O. The structure of butanal is shown below:

Use ideas about bonds to explain whether the enthalpy change of combustion of butanal is likely to be similar to that of butanone. (*3 marks*)

5.1 Ionic substances in solution

Specification reference: EL (s)

Many ionic substances dissolve well in water – they are described as being soluble compounds.

- All compounds of Group 1 metals are soluble.
- All compounds containing nitrate ions are soluble.
- All compounds containing ammonium ions are soluble.

Insoluble compounds

Some ionic substances do not dissolve well in water – they are described as insoluble compounds:

- sulfates of barium, calcium, lead, and silver
- halides (chlorides, bromides, and iodides) of silver and lead
- all carbonates (except those of Group 1 ions or ammonium ions)
- hydroxides containing some Group 2 ions, aluminium ions, or d-block ions.

Forming precipitates

Certain combinations of ions form insoluble compounds (see above).

If solutions containing these ions are mixed together, a precipitate will form.

Challenge

Almost all substances are soluble to some extent; for example, the solubility of magnesium hydroxide is $2.00 \times 10^{-4} \, mol \, dm^{-3}$

a Convert this into a concentration in g / 100 cm^3 of solution.

b A substance is sometimes described as insoluble if the solubility is less than 0.1 g / 100 cm^3 solution. Should magnesium hydroxide be classified as soluble or insoluble?

Precipitates containing d-block ions may have particular colours.

Typical results for combinations of cations and anions are shown below (ppt = precipitate), using solutions with a concentration of around $0.1 \, mol \, dm^{-3}$.

Some of the reactions are beyond the scope of the course and have been omitted.

Key term

Precipitate: A suspension of solid particles formed by a chemical reaction in solution.

Revision tip

You need to learn the colours of the coloured precipitates in this table.

Synoptic link

You will need to be able to explain how to use a sequence of these tests to identify salts. See Experimental techniques.

	OH^-	SO_4^{2-}	CO_3^{2-}	Cl^-	Br^-	I^-
Ca^{2+}	White ppt	White ppt	White ppt	Soluble	Soluble	Soluble
Ba^{2+}	White ppt	White ppt	White ppt	Soluble	Soluble	Soluble
Cu^{2+}	Pale blue ppt	Soluble	Green ppt	Soluble	Soluble	Soluble
Fe^{2+}	Green ppt	Soluble	Green ppt	Soluble	Soluble	Soluble
Fe^{3+}	Brown ppt	Soluble	–	Soluble	Soluble	Soluble
Al^{3+}	White ppt	Soluble	–	Soluble	Soluble	Soluble
Pb^{2+}	White ppt	White ppt	White ppt	White ppt	White ppt	Yellow ppt
Zn^{2+}	White ppt	Soluble	White ppt	Soluble	Soluble	Soluble
Ag^+	White ppt	White ppt	White ppt	White ppt	Cream ppt	Yellow ppt

Synoptic link

Precipitation reactions like these are often described using ionic equations. See Topic 1.2, Balanced equations.

Synoptic link

Precipitation reactions involving halide ions (including the solubility of the precipitates in ammonia) are described in Topic 11.3, The p-block: Group 7.

Testing for ions

The formation of precipitates can be used as a test for certain ions:

- Adding barium chloride solution (containing Ba^{2+} ions): white precipitate shows that sulfate ions are present.

- Adding silver nitrate solution (containing Ag^+ ions): a precipitate shows that halide ions are present (white = chloride, cream = bromide, yellow = iodide).

Summary questions

1 Describe how you would carry out a test to show the presence of:
   a bromide ions   (2 marks)
   b sulfate ions.   (2 marks)

2 What colours are the precipitates formed from the combinations of the following ions:
   a copper(II) ions and hydroxide ions
   b calcium ions and carbonate ions
   c iron(III) ions and hydroxide ions
   d silver ions and iodide ions.   (2 marks each)

3 A pale green solution solution containing an ionic salt, A, was investigated using precipitation reactions. When sodium hydroxide solution was added to the solution, a green precipitate was seen. When barium chloride solution was added a white precipitate was formed:
   a identify substance A   (2 marks)
   b write ionic equations for the reactions that form the precipitates in the two tests.   (2 marks)

5.2 Bonding, structure, and properties

Specification reference: EL (j), EL (l)

Structure and bonding

Types of structure

The types of structure present in a substance depends on the type of bonding present.

Type of bonding	Type of structure	Example
Ionic	Giant ionic	NaCl
Covalent	Simple molecular	CO_2
Covalent	Giant covalent network	SiO_2
Metallic	Giant metallic lattice	Fe

Giant ionic lattice

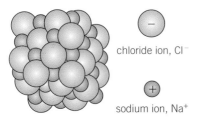

chloride ion, Cl^-

sodium ion, Na^+

▲ **Figure 1** *Structure of a sodium chloride lattice*

The structure of a giant ionic lattice

There is a regular repeating pattern of positive and negatively charged ions in all three dimensions. The attraction between these oppositely charged ions outweighs the repulsion between ions with the same charge, because the oppositely charged ions are closer.

The sodium chloride structure

You can see in the diagram above that chloride, Cl^- ions are larger than sodium ions, Na^+. The ions are arranged so that 6 Cl^- ions surround 1 Na^+ ion in a 3-dimensional arrangement.

The ratio of Cl^- to Na^+ is 1:1, so 6 Na^+ also surround 1 Cl^-.

Characteristic properties

- High melting point, because there are strong electrostatic attractions between ions
- Often soluble in water
- Conduct electricity when molten or in solution, because the charged ions are able to move in response to a voltage.

Simple molecular structure

There are strong covalent bonds within the molecules, but only weak intermolecular bonds between the molecules.

Characteristic properties

- Low melting point
- Usually insoluble in water
- Do not conduct.

Synoptic link

A description of the electrostatic forces in ionic and covalent bonds is found in Topic 3.1, Chemical bonding.

chloride ion Cl^- sodium ion Na^+

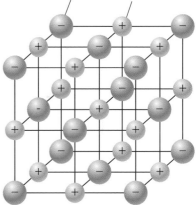

▲ **Figure 2** *An alternative way of representing the lattice to show the 3-dimensional arrangement of ions*

Challenge

The 6 Cl^- ions will arrange themselves to be as far apart as possible. Suggest the name given to the 3-dimensional arrangement that they will take up.

Synoptic link

Intermolecular bonds and the properties of molecular substances are explained in Topics 5.3 and 5.4.

Giant covalent network

Some covalent substances have a giant network structure. There are strong covalent bonds between the atoms in the network.

Characteristic properties

- High melting point, because all the bonds in the structure are strong covalent bonds
- Insoluble in water
- Do not conduct electricity (except for graphite).

▲ **Figure 3** *Diamond has a giant covalent network structure*

Common misconception: Molecules or giant structures?

Silicon dioxide, SiO_2 has a giant structure, but does not consist of separate SiO_2 molecules. The giant structure simply contains Si and O atoms in the ratio 1:2, all strongly bonded together in a 3-dimensional lattice.

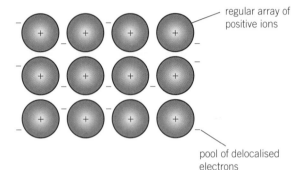

regular array of positive ions

pool of delocalised electrons

▲ **Figure 4** *A giant metallic lattice structure*

Giant metallic lattice

All metals have a giant metallic lattice structure. There is a strong electrostatic attraction between the positive metal ions and the delocalised electrons between the ions.

Characteristic properties

- High melting point, because there is strong attraction between ions and electrons
- Insoluble in water
- Conduct electricity when solid or molten because the electrons are free to move in response to a voltage.

Key term

Delocalised electrons: Electrons that are not associated with a particular atom or bond, but are free to move over several atoms.

Challenge

Group 1 metals have relatively low melting points. Use ideas about the charge and size of Group 1 ions to explain this.

Summary questions

1 Describe the structure and bonding in the metal calcium. *(4 marks)*

2 Calcium oxide, $CaCl_2$ has a giant ionic structure and sulfur dichloride, SCl_2 has a simple molecular structure. Describe the differences in properties between these substances. *(3 marks)*

3 Silicon dioxide is a high melting point solid, but carbon dioxide is a gas at room temperature. Use ideas about structure and bonding to explain this difference in properties. *(5 marks)*

5.3 Bonds between molecules: temporary and permanent dipoles

Specification reference: OZ (a), OZ (b)

Bonds between molecules

In any liquid or solid, there are bonds between molecules. These are called intermolecular bonds. One type of intermolecular bond is a dipole–dipole bond. Dipole–dipole bonds can be:

- permanent dipole–permanent dipole bonds
- instantaneous dipole–induced dipole bonds.

Electronegativity and dipoles

Permanent dipoles can arise in molecules, because some molecules contain polar bonds:

- Covalent bonds can be polar if the two atoms in the bond have different electronegativities.
- The charges in a polar bond are shown using δ^+ and δ^- symbols.

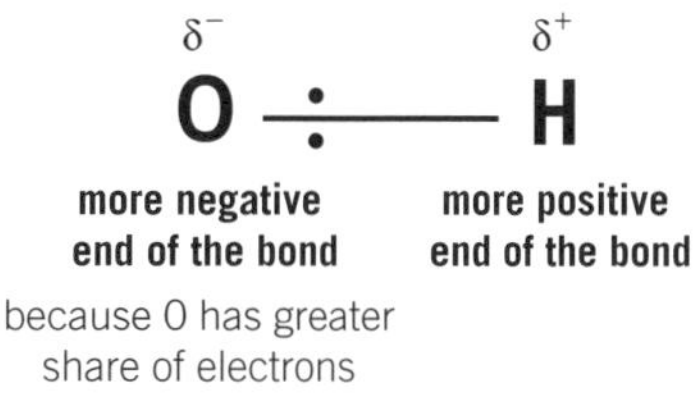

$$\overset{\delta^-}{O} \;\; \vdots \;\; \overset{\delta^+}{H}$$

more negative end of the bond   **more positive end of the bond**

because O has greater share of electrons

▲ **Figure 1** *A polar O−H bond*

Electronegativity values

Electronegativity is related to the position of an atom in the periodic table.

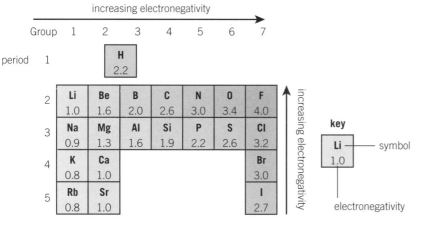

▲ **Figure 2** *Trends in electronegativity in the periodic table*

In a bond between two atoms, the more electronegative atom has a δ^- charge and the less electronegative has a δ^+ charge.

Dipoles in molecules

Molecules that contain polar bonds may have a permanent dipole (also called an overall dipole). They are also sometimes called polar molecules.

> 🖩 **Worked example: Overall dipole of CCl$_4$**
>
> The molecule tetrachloromethane, CCl_4, has a tetrahedral structure. Does it have an overall (permanent) dipole?
>
> **Step 1:** Decide whether there are any polar bonds in the molecule:
>   *C−Cl is polar because Cl is more electronegative than C*

$$\overset{\delta+}{\cdots H} \overset{\delta-}{-Br} \overset{\delta+}{\cdots H} \overset{\delta-}{-Br} \overset{\delta+}{\cdots H} \overset{\delta-}{-Br}$$

▲ **Figure 3** *Permanent dipole–permanent dipole bonds between HBr molecules*

Common misconception: Dipole–dipole attractions

If a molecule has an overall dipole, then there will be two types of intermolecular bonds: permanent dipole–permanent dipole bonds and instantaneous dipole–induced dipole bonds. There may even be hydrogen bonds as well (see Topic 5.4, Bonds between molecules).

Synoptic link

You need to be able to apply these ideas to explain the patterns in boiling point of alkanes (Topic 12.1, Alkanes) and of the halogens (Topic 11.3, The p-block: Group 7).

Longer chains increase the number of instantaneous dipole-induced dipole bonds

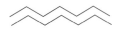

octane: T_b: 399 K

pentane: T_b: 309 K

Increased branching decreases the surface area in contact and decreases the strength of instantaneous dipole-induced dipole bonds

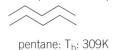

pentane: T_b: 309K

methyl butane: T_b: 301K

Molecules with more electrons have more chance of forming instantaneous dipoles, and the strength of instantaneous-dipole induced dipole bonds increases

Cl—Cl	Br—Br	I—I
34 electrons	70 electrons	106 electrons
T_b: 238K	T_b: 332 K	T_b: 457 K

▲ **Figure 4** *The factors that affect instantaneous dipole-induced dipole bonds can be used to explain the boiling points of alkanes and halogen molecules*

◀ **Step 2:** If there are any polar bonds, draw out the 3-dimensional shape of the molecule, marking on the partial charges:

tetrachloromethane

Step 3: Decide whether the charges are arranged symmetrically – in other words, whether the centre of positive charge is in the same place as the centre of negative charge:

The charges are arranged symmetrically, so there is no overall dipole in a CCl_4 molecule

Permanent dipole–permanent dipole bonds

If two neighbouring molecules both have a permanent dipole, then there will be an electrostatic attraction between the charges of these dipoles. This is called a permanent dipole–permanent dipole bond.

Instantaneous dipole–induced dipole bonds

If two neighbouring molecules do not have a permanent dipole, there will still be an attraction between the molecules. This is called an instantaneous dipole–induced dipole bond. Instantaneous dipole–induced dipole attraction occurs between molecules, even if permanent dipoles are also present.

Explaining how these bonds arise

- Electrons in a molecule are in continuous random motion.
- At a particular instant, the electrons may be distributed unevenly. This creates an instantaneous dipole.
- The dipole induces a dipole on a neighbouring molecule, creating an induced dipole.
- There is an electrostatic attraction between the two dipoles.

Because the electron distribution in a molecule is constantly changing, instantaneous dipole–induced dipole bonds are continually breaking and reforming.

Factors affecting strength of instantaneous dipole–induced dipole bonds

- Number of electrons in the molecule: more electrons mean more chance of an instantaneous dipole arising
- Distance between the molecules – closer packing means a greater electrostatic attraction.

Summary questions

1 Show the partial charges that exist on the atoms in these bonds:
   a C–Cl, **b** Cl–F, **c** H–N, **d** Cl–S *(4 marks)*

2 CCl_4 does not have an overall dipole:
   a State the type of intermolecular bonds between two molecules of CCl_4. *(1 mark)*
   b Explain how these bonds arise. *(5 marks)*

3 The molecule CCl_2F_2 contains polar bonds. Discuss whether the molecule has an overall dipole. *(4 marks)*

5.4 Bonds between molecules: hydrogen bonding

Specification reference: OZ (c), OZ (d)

Hydrogen bonds

In general, hydrogen bonds are the strongest type of intermolecular bond.

Requirements for a hydrogen bond between two molecules

A δ^+ H atom in one molecule (e.g. a H atom bonded to an electronegative atom such as O, N, or F). A small electronegative atom (O, N, or F) in the other molecule. A lone pair on the electronegative atom.

Arrangement of atoms in a hydrogen bond

The lone pair points directly at the δ^+ H atom. The arrangement of atoms around the δ^+ H is linear, so the bond angle is 180°.

Hydrogen bonds in water and ice

Liquid water

Water molecules contain an O atom with 2 lone pairs. There are 2 δ^+ H atoms in each water molecule. In a collection of several water molecules, each water molecule can, on average, form up to *two* hydrogen bonds with neighbouring molecules. Liquid water has a higher boiling point than other small molecules with similar M_r values, such as CH_4 or NH_3.

Ice

The hydrogen bonds formed when water freezes give ice a regular structure. The hydrogen bonds and covalent bonds around each O atom are arranged tetrahedrally. This arrangement of bonds around the O atom gives ice a very open structure. As a result, it has a lower density than water. So ice floats on water.

Comparing boiling points of molecules

Ideas about intermolecular bonds can be used to compare the boiling points of two substances. The stronger the intermolecular bonds, the higher the boiling point.

Summary questions

1   A hydrogen bond can form between two molecules of ammonia, NH_3. Explain why.   (3 marks)
2   Draw a diagram showing how a hydrogen bond can form between two molecules of ethanol. Show partial charges and relevant lone pairs.   (4 marks)

3   The boiling points of water, hydrogen fluoride, and methane are: 373 K, 293 K, and 109 K. Explain the differences in boiling points of these molecules.   (4 marks)

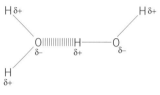

▲ **Figure 1** *In a collection of water molecules there will be 2 hydrogen bonds per molecule*

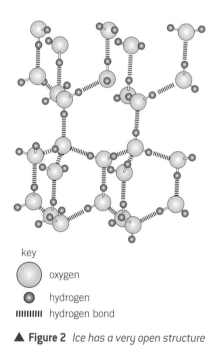

key

○ oxygen

● hydrogen

|||||||| hydrogen bond

▲ **Figure 2** *Ice has a very open structure*

Analysing an unknown salt X

You need to be able to describe how to carry out tests to identify a range of possible cations and anions present in a sample of an unknown salt, X.

Testing for anions

You can use simple test tube reactions to test for the presence of some common negative ions in an unknown compound X.

Carbonate

Carbonates react with acid to form carbon dioxide gas:

- In a test tube, add dilute acid to a solid or solution of X.
- If bubbles of gas are formed, then the solid probably contains a carbonate (or hydrogen carbonate) ion.
- To check that the gas is carbon dioxide, bubble the gas through limewater (saturated calcium hydroxide solution). A milky-white precipitate will be formed.

Sulfate

Barium sulfate is insoluble:

- Add barium chloride (or barium nitrate) solution to a solution of X.
- A white precipitate indicates the presence of sulfate ions.

Halide ions

Silver halides are insoluble and have noticeably different colours:

- Add silver nitrate solution (acidified with a few drops of nitric acid) to a solution of X.
- A precipitate indicates the presence of halide ions.
- Chloride ions produce a white precipitate, bromide ions produce a cream precipitate, and iodide ions a yellow precipitate.

Testing for metal cations

Precipitation reactions

Some metal cations produce coloured precipitate with sodium hydroxide, which can be used to identify the cation. Add sodium hydroxide solution to a solution of X. Observe the colour of any precipitate.

Flame tests

A piece of nichrome wire is cleaned in concentrated hydrochloric acid and dipped into a solution of X or a powdered sample of X. The end of the wire is heated in a blue Bunsen flame. Different metal ions cause different colours to be observed in the flame.

Revision tip

Sequence of tests for anions

When testing an unknown solution for a range of anions, the tests should be done in a particular order:

1 Test for carbonate ions using nitric acid as the acid.

2 Test for sulfate ions; filter off any precipitate formed.

3 Test for halide ions.

Cation present	Observation
Cu^{2+}	Blue precipitate
Fe^{2+}	Green precipitate
Fe^{3+}	Brown precipitate
Ca^{2+}, Ba^{2+}, Al^{3+}, Pb^{2+}, Zn^{2+}	White precipitate

Revision tip

If no precipitate is observed in this test, then X probably contains a Group 1 ion or an ammonium ion.

Synoptic link

You need to be able to use ideas about electron energy levels to describe why metal ions emit visible light. See Topic 6.1, Light and electrons.

Chapter 5 Practice questions

1   Which combination of ions will produce a precipitate?

    1 Ba^{2+} ions and Cl^- ions

    2 Fe^{2+} ions and OH^- ions

    3 Ag^+ ions and I^- ions

  A 1, 2, and 3

  B Only 1 and 2

  C Only 2 and 3

  D Only 3   *(1 mark)*

2   Which of these options is the best description of the likely properties of the compound boron trifluoride, BF_3?

  A High melting point, insoluble in water, conducts electricity when molten.

  B Low melting point, insoluble in water, does not conduct electricity when molten.

  C Low melting point, soluble in water, conducts electricity when molten.

  D High melting point, insoluble in water, does not conduct electricity when molten.   *(1 mark)*

3   Which of these molecules has an overall dipole?

    1 CCl_2H_2

    2 CCl_3F

    3 CCl_4

  A 1, 2, and 3

  B Only 1 and 2

  C Only 2 and 3

  D Only 3   *(1 mark)*

4   Ice has a lower density than water. The best explanation of this is:

  A Hydrogen bonds are longer in ice than in liquid water.

  B Ice contains pockets of air that are trapped by the hydrogen bonds in the structure.

  C The tetrahedral arrangement of bonds and hydrogen bonds around each O atom holds the molecules in an open structure.

  D The hydrogen bonds cause the covalent bonds in the water molecules to lengthen so the molecules take up more space. *(1 mark)*

5   Propanone, butane, propan-1-ol, and ethane-1,2-diol all have similar molecular masses. The most likely order of the boiling points of these molecules, from low to high, is:

  A propanone, butane, ethane-1,2-diol, propan-1-ol

  B ethane-1,2-diol, propanone, propanol, butane

  C propanone, butane, propanol, ethane-1,2-diol

  D butane, propanone, propan-1-ol, ethane-1,2-diol *(1 mark)*

6   A solution of an ionic salt, X is investigated using a series of tests. No precipitate is observed when barium chloride is added, but a cream precipitate is seen when acidified silver nitrate is added.

Sodium hydroxide solution was added to a second sample and a brown precipitate was seen. The most likely identity of X is:

A Iron(III) bromide

B Iron(II) sulfate

C Copper(II) chloride

D Iron(III) sulfate   (*1 mark*)

7 The concentration of carbon dioxide in the troposphere is 0.4%, whereas the concentration of carbon monoxide is 0.1 ppm. How many times greater is the carbon dioxide concentration than the carbon monoxide concentration?

A 400

B 40 000

C 3.9

D 4 000 000   (*1 mark*)

8 During the second half of the twentieth century, a class of synthetic molecules known as CFCs (chlorofluorocarbons) were used in large quantities in several different applications.

These molecules are relatively volatile and the widespread use of CFCs resulted in large amounts of these molecules being released into the atmosphere.

   a One such CFC molecule is CCl_3F. The structure of this molecule is shown below:

The C–Cl and C–F bonds in this molecule are polar and the molecule itself has an overall dipole

   i   State the meaning of the word polar when used to describe bonds   (*1 mark*)

   ii Explain why the bonds in this CFC molecule are polar   (*2 marks*)

   iii Explain why CCl_3F has an overall dipole   (*2 marks*)

   iv Name the strongest type of intermolecular bond that exists between two molecules of CCl_3F.   (*1 mark*)

   b A related molecule, CCl_4 has a similar structure to CCl_3F, but a Cl atom replaces the F atom. The strongest intermolecular bonds between two molecules of CCl_4 are instantaneous dipole–induced dipole bonds.

   i   Describe how these bonds arise.   (*4 marks*)

   ii CCl_4 has a higher boiling point than CCl_3F. Discuss the factors that account for this difference   (*5 marks*)

6.1 Light and electrons

Specification reference: EL (w), EL (v), OZ (u)

Bohr's theory and wave-particle duality

Bohr model of the atom

Electrons in an atom occupy shells, or energy levels. Electrons in an energy level have a specific amount of energy; the energy of electrons is said to be quantised.

Atomic emission spectra

Appearance of an emission spectrum

An emission spectrum is seen when atoms are heated up to a high temperature. It consists of coloured lines on a black background:

- The lines become closer at higher frequencies.
- There are several series of lines (although some of these may fall outside the visible part of the spectrum).

Explaining the formation of an emission spectrum

- Electrons in the ground state absorb energy.
- This energy causes electrons to be excited to a higher electron energy level.
- Electrons then drop back to lower energy levels. The energy lost (ΔE) is emitted as a photon of light.
- The light emitted is seen as a specific line in the spectrum.
- The frequency of the photon is related to the energy lost by the equation: $\Delta E = h\nu$.
- Different energy gaps produce photons of different frequencies.

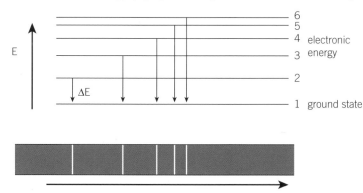

▲ **Figure 1** *An emission spectrum*

Flame colours

Different metal ions produce different flame colours (see Table 1).

▼ **Table 1**

Ion	Colour
Li^+	Bright red
Na^+	Yellow
K^+	Lilac
Ca^{2+}	Orange-red
Ba^{2+}	Pale green
Cu^{2+}	Blue-green

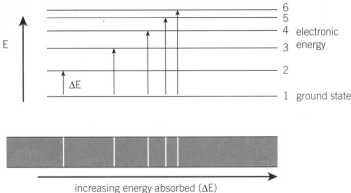

increasing energy absorbed (ΔE)
increasing frequency of light absorbed

▲ **Figure 2** *The formation of an atomic absorption spectrum*

Atomic absorption spectra
Appearance of an absorption spectrum
If white light is passed through a sample containing vaporised atoms, an atomic absorption spectrum is seen.

Formation of an absorption spectrum
- Electrons in the ground state absorb photons of light
- This energy from these photons causes electrons to be excited to a higher electron energy level
- The frequency of the photon is related to the energy gained by the equation: $\Delta E = h \nu$
- Light of this frequency does not pass through the sample, so a black line is seen in the spectrum.

Similarities and differences between the two types of spectrum
Similarities
- For a given element, lines appear at the same frequency
- Lines converge at a higher frequency
- Several series of lines are seen.

Differences
- Emission spectrum: coloured lines on a black background
- Absorption spectrum: black lines on a coloured background.

Calculations using data from lines in the spectrum
You need to be able to link energy (of the photon absorbed or emitted), frequency and wavelength. The key equations are:

$$\Delta E = h \nu \text{ and } c = \nu \lambda.$$

ν = frequency in Hz
h = Planck constant, $6.63 \times 10^{-34}\,\text{J}\,\text{Hz}^{-1}$
c = speed of light, $3 \times 10^{8}\,\text{m}\,\text{s}^{-1}$
λ = wavelength in m.

Revision tip
When you are doing these calculations, remember that ΔE refers to an electron transition in a single atom, so the value of ΔE, in J, will be very small.

Summary questions

1 State the flame colour associated with:
  a sodium ions
  b barium ions. (*2 marks*)

2 Discuss the difference between an atomic emission spectrum and an atomic absorption spectrum. You should include comments about the appearance of the spectra and the processes in atoms that produce the spectra.
  (*4 marks*)

3 The atomic emission spectrum of a hydrogen atom consists of several series of lines. Explain how these series of lines are formed. (*5 marks*)

🖩 **Worked example: Calculations using the wavelength of light**

The emission spectrum of sodium contains a line with $\lambda = 590$ nm [1 nm $= 10^{-9}$ m]. Calculate the change in energy of the electronic transition that causes this line in the spectrum.

Step 1: Convert the wavelength to m: 590 nm $= 590 \times 10^{-9} = 5.90 \times 10^{-7}$ m.

Step 2: Calculate the frequency of this light: $c = \nu \lambda$, so $\nu = c/\lambda = \frac{3.00 \times 10^{8}}{5.90 \times 10^{-7}}$
$= 5.085 \times 10^{14}$ Hz

Step 3: Calculate the energy change that produces light of this frequency: $\Delta E = h \nu$,
so $\Delta E = 6.63 \times 10^{-34} \times 5.085 \times 10^{14} = 3.37 \times 10^{-19}$ J.

6.2 What happens when radiation interacts with matter?

Specification reference: EL (v), OZ (s), OZ (t), OZ (u)

The electromagnetic spectrum

Visible light, ultraviolet radiation, and infrared radiation are all examples of electromagnetic radiation.

	radiofrequency		micro-wave	infrared	visible	UV	X-rays	γ-rays
Frequency (Hz) 10^5 10^6 10^7 10^8 10^9			10^{10} 10^{11}	10^{12} 10^{13}	10^{14} 10^{15}	10^{16} 10^{17}	10^{18} 10^{19} 10^{20}	
Wavelength (m) 10^3		1		10^{-3}	10^{-6}		10^{-9}	

▲ **Figure 1** *The electromagnetic spectrum, showing the wavelengths and frequencies of the different types of radiation*

Radiation from the Earth and the Sun

The surface of the Earth and the surface of the Sun have very different temperatures. They both emit electromagnetic radiation, but of very different types.

The effect of radiation on matter

Electromagnetic radiation can interact with matter, transferring energy to the atoms or bonds present. The effect that this has on matter depends on the type of radiation involved.

Type of radiation	Effect on matter
Infrared	Bonds vibrate with greater energy
Visible	Electrons excited to higher energy levels; some bonds can break
Ultraviolet	Electrons excited to higher energy levels and bonds break

Calculations to do with bond breaking

You can use information about the bond enthalpy of covalent bonds to calculate the minimum energy needed to cause a bond to break.

> **Worked example: Calculating the frequency needed to cause photodissociation**
>
> Electromagnetic radiation can cause the photodissociation of bromomethane, CH_3Br in the Earth's atmosphere. The C−Br bond has a bond enthalpy of 276 kJ mol⁻¹. Calculate the minimum frequency of radiation needed to break the C−Br bond.
>
> **Step 1:** Convert the bond enthalpy to J: 276 kJ mol⁻¹ = 276 000 J mol⁻¹
>
> **Step 2:** This is the energy needed for 1 mole of bonds (= N_A bonds, where N_A = Avogadro constant). Calculate the energy needed to break 1 bond:
> $$N_A = 6.02 \times 10^{23} \text{ mol}^{-1}.$$
> So energy needed to break 1 bond = $276\,000/6.02 \times 10^{23} = 4.585 \times 10^{-19}$ J.
>
> **Step 3:** Calculate the frequency of a photon with this energy, using $E = h\nu$:
> $$\nu = \frac{E}{h} = \frac{4.585 \times 10^{-19}}{6.63 \times 10^{-34}} = 6.92 \times 10^{14} \, Hz$$

Summary questions

1 Name the principal type of electromagnetic radiation emitted by:
   a the surface of the Earth
   b the Sun   (*2 marks*)

2 Visible light and infrared radiation differ from each other in terms of:
   a the properties of the photons, and
   b the effect they have on matter. Describe these differences. (*5 marks*)

3 The minimum frequency of radiation needed to break one C−H bond is 1.09 × 10^{15} Hz. Calculate the bond enthalpy of a C−H bond, in kJ mol⁻¹ (*3 marks*)

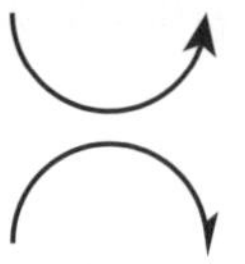

▲ **Figure 1** *A full curly arrow (top) and a half curly arrow (bottom)*

Synoptic link

You need to be able to use full curly arrows to show the movement of a pair of electrons in organic mechanisms. See Topic 12.2, Alkenes and 13.2, Haloalkanes.

Key terms

Heterolytic fission: A type of covalent bond breaking, in which both electrons from the shared pair go to just one atom.

Homolytic fission: A type of covalent bond breaking, in which one electron from the shared pair goes to each atom.

Radical: A species with one (or more) unpaired electrons.

Key term

Radical chain reaction: A reaction in which new radicals are formed at the end of one step, which continue. These radicals then continue the process.

Revision tip

Dots are not always included in the formulae of radicals. You may need to deduce whether a species is a radical by counting the total number of outer shell electrons: if there is an odd number, then the species must be a radical. Some species (e.g. O atoms) have two unpaired electrons and are called biradicals. Dots are not usually used for biradicals.

Bond fission

Covalent bonds can break in two different ways, producing either ions or radicals:

$$H \overset{\frown}{\underset{\bullet}{\bullet}} Cl \longrightarrow H^+ + Cl^-$$

▲ **Figure 2** *Heterolytic fission, producing ions*

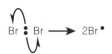

▲ **Figure 3** *Heterolytic fission, producing radicals*

Curly arrows

These are used to show the movement of electrons.

● A half curly arrow shows the movement of one electron.
● A full curly arrow shows the movement of a pair of electrons.

Radicals and chain reactions

Radicals are highly reactive and undergo chain reactions. These reactions are often very fast, and may occur in the presence of light. You can often recognise radicals in equations, because they are written with a dot to indicate the unpaired electron:

$$Cl\bullet \qquad NO\bullet \qquad CH_3\bullet$$

Mechanism of a chain reaction

Radical chain reactions happen in three stages:

1 Initiation: radicals are formed from a stable molecule.

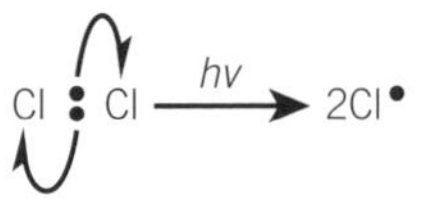

$$Cl \overset{\bullet}{\underset{\bullet}{\bullet}} Cl \xrightarrow{\ h\nu\ } 2Cl^\bullet$$

2 Propagation: a radical reacts and the process forms a new radical.

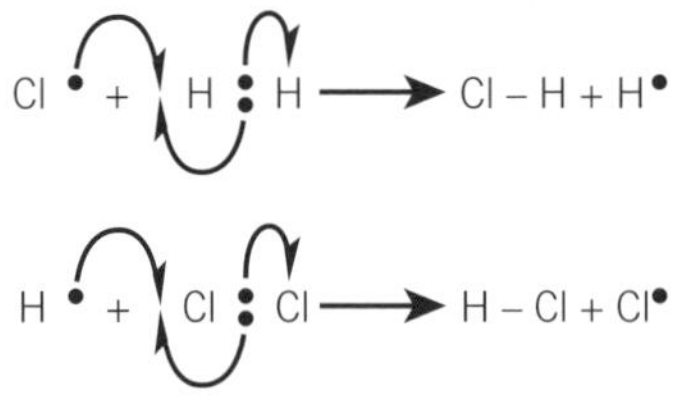

$$Cl^\bullet + H \overset{\bullet}{\underset{\bullet}{\bullet}} H \longrightarrow Cl-H + H^\bullet$$

$$H^\bullet + Cl \overset{\bullet}{\underset{\bullet}{\bullet}} Cl \longrightarrow H-Cl + Cl^\bullet$$

Propagation steps often occur in pairs; the radical formed by the 1st step reacts again in the 2nd step.

3 Termination: two radicals collide to form a stable molecule.

$$H^\bullet + H^\bullet \longrightarrow H-H$$

Examples of radical chain reactions
Reactions of alkanes with halogens

A halogen atom can substitute for a hydrogen atom in an alkane chain. The mechanism of this reaction is radical substitution.

1. Initiation: Homolytic fission of a halogen molecule in the presence of UV light:

$$Cl_2 + h\nu \rightarrow 2Cl\bullet$$

2. Propagation steps: A methyl radical is formed, which then reacts to reform the Cl•.

The Cl• can be thought of as a catalyst in these stages:

$$CH_4 + Cl\bullet \rightarrow CH_3\bullet + HCl$$

$$CH_3\bullet + Cl_2 \rightarrow CH_3Cl + Cl\bullet$$

3. Termination: Cl• or CH₃• radicals collide:

$$Cl\bullet + Cl\bullet \rightarrow Cl_2$$

$$CH_3\bullet + Cl\bullet \rightarrow CH_3Cl$$

The depletion of ozone by Cl atoms

Ozone, O_3 is present in the stratosphere. Halogen atoms from synthetic compounds such as haloalkanes act as catalysts for the breakdown of ozone. (In the mechanism below, the dots showing unpaired electrons have not been included.):

1. Chloroalkanes reach the stratosphere and photodissociate, forming Cl:

$$CH_3Cl + h\nu \rightarrow CH_3\bullet + Cl\bullet$$

2. The chlorine reacts with ozone in a catalytic cycle, involving O atoms, which are also present in the stratosphere:

$$Cl\bullet + O_3 \rightarrow ClO\bullet + O_2$$

$$ClO\bullet + O \rightarrow Cl\bullet + O_2$$

3. Chlorine atoms are removed from the cycle by termination reactions:

$$Cl\bullet + Cl \rightarrow Cl_2$$

Challenge

The overall effect of a 2-stage process can be deduced by adding together the reactants and products of both stages and then cancelling out any species that appear on both sides. Write overall equations for the propagation steps in:

a   the reaction of methane with chlorine

b   the depletion of ozone by chlorine radicals.

Photodissociation of other haloalkanes

Iodoalkanes and bromoalkanes photodissociate more easily than chloroalkanes, because C−I and C−Br bonds are weaker than C−Cl bonds. This means that they can be broken down by the lower frequency radiation in the troposphere and do not reach the stratosphere. Fluoroalkanes do not photodissociate in the stratosphere, because the C−F bond is too strong to be broken by the ultraviolet radiation present.

The importance of ozone

Ozone in the stratosphere

Ozone absorbs high-energy ultraviolet radiation from the Sun. It prevents this radiation reaching the surface of the Earth. High-energy ultraviolet radiation can cause health problems, including skin cancer and eye cataracts.

Ozone in the troposphere

Ground level ozone is a secondary pollutant, present as a component of photochemical smog.

Revision tip

The presence of ultraviolet radiation (or visible light) in a reaction is sometimes shown using the symbol 'hv' in the equation.

Synoptic link

You may need to be able to compare radical substitution with other substitution reactions, such as nucleophilic substitution in Topic 13.2, Haloalkanes.

Revision tip

Ozone can also be depleted by the action of other radicals present in the stratosphere, such as NO and OH. You should be able to deduce the equations for depletion processes involving these radicals.

Synoptic link

You also need to be able to link the trend in bond enthalpy of C-halogen bonds to the reactivity of haloalkanes with nucleophiles. This is covered in Topic 13.2, Haloalkanes.

Key terms

Troposphere: The layer of the atmosphere immediately above the Earth's surface.

Stratosphere: The layer of the atmosphere above the troposphere.

Synoptic link

Photochemical smog is formed from a reaction between hydrocarbons and nitrogen oxides. The formation of these pollutants is described in Topic 15.1, Atmospheric pollutants. Photochemical smog causes breathing problems and the corrosion of plastics, rubber, and textiles.

The formation and destruction of ozone

Ozone formation

Ozone is formed naturally in the stratosphere:

1   Oxygen molecules photodissociate into oxygen atoms:

$$O_2 + h\nu \rightarrow 2O$$

2   Ozone is formed when an oxygen atom combines with an oxygen molecule:

$$O_2 + O \rightarrow O_3$$

Ozone destruction

Ozone is destroyed naturally when it absorbs high energy ultraviolet radiation:

$$O_3 + h\nu \rightarrow O_2 + O$$

Summary questions

1   State two differences between homolytic and heterolytic bond fission. *(2 marks)*

2   Ozone molecules in the stratosphere absorb high energy ultraviolet light:
    a   describe the chemical process that occurs when this happens
    b   explain why this process is important for human health at the surface of the Earth. *(3 marks)*

3   NO radicals can deplete ozone in a similar way to the action of Cl radicals. Write equations for a catalytic cycle involving NO and ozone molecules. *(2 marks)*

6.4 Infrared spectroscopy

Specification reference: WM (j)

The effect of infrared radiation on molecules

Each type of covalent bond in a molecule vibrates at a specific frequency. When a bond absorbs infrared radiation of the correct frequency, it vibrates with greater energy. So molecules may absorb infrared radiation at several different frequencies, depending on the bonds that are present.

The appearance of an infrared spectrum

An infrared spectrum shows the % transmittance of infrared over a range of wavenumbers.

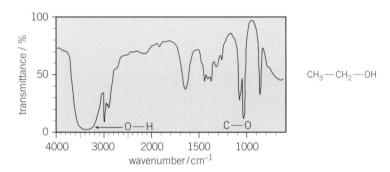

▲ **Figure 1** *The infrared spectrum of ethanol*

- Absorption of infrared radiation is shown by the downward pointing troughs – called peaks – in the spectrum.
- The strength and shape of the peak may also be helpful when interpreting the spectrum – peaks may be strong or broad.

> ### Model answer: Explain how you can tell that spectrum A is the spectrum of an alcohol, and not the spectrum of a carboxylic acid.
>
> In spectrum A there is a broad peak at 3200–3600 cm^{-1}, indicating an O–H in an alcohol group; if the O–H had been part of a carboxylic acid group, it would be very broad and would extend between 2500 and 3200 cm^{-1}. There is no peak at 1700–1725 cm^{-1}, so there is no C=O bond in a carboxylic acid group.

Fingerprint region

The peaks above 1500 cm^{-1} help you to identify the types of bond in the molecule. The pattern of peaks below 1500 cm^{-1} is very complex and difficult to analyse. This fingerprint region is unique to a specific molecule and so, by comparing it to a computer database, the exact identity of a molecule can be found.

Summary questions

1. Describe what happens in a molecule when it absorbs infrared radiation. *(2 marks)*

2. The infrared spectrum of ethanoic acid contains a broad peak between 2 500 and 3 200 cm^{-1} and a sharp peak at 1 715 cm^{-1}. Explain why ethanoic acid has peaks at these wavenumbers. *(4 marks)*

3. A C–H bond vibrates at a frequency of 9.04×10^{13} Hz. Calculate the wavenumber of infrared radiation that will be absorbed by this C–H bond. *(3 marks)*

> ### Revision tip
> When describing or interpreting a spectrum, you refer to the bond, the location (the functional group containing the bond), and wavenumber range.

> ### Revision tip
> In exam answers, you should always discuss regions 3200–3600 cm^{-1} (where O–H bonds show up) and 1700–1750 cm^{-1} (where most C=O bonds show up). The absence of a peak may be as important as its presence.

> ### Common misconception: C–H bonds
> C–H bonds are present in all molecules, so infrared spectra of organic molecules generally have a peak or set of peaks at around 3000 cm^{-1}. However, they are not often useful in helping to identify functional groups in molecules, except in some cases, for example alkenes or arenes.

> ### Key term
> **Fingerprint region:** The region in an infrared spectrum below 1500 cm^{-1}.

A mass spectrometer is an instrument that measures the masses and abundances of positive (1+) ions produced from atoms and molecules.

You need to know how to interpret two types of mass spectra.

- the mass spectrum produced from a sample of an element
- the mass spectrum produced from a sample of a molecule.

Mass spectra of elements

A sample of an element may contain several isotopes, with different abundances.

Deductions from the mass spectrum of an element

From the mass spectrum of an element you can deduce:

- the number of isotopes in the sample from the number of peaks in the spectrum
- the relative isotopic masses of each isotope from the m/z values
- the % abundance of each isotope from the height of the peaks.

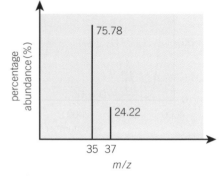

Mass spectra of molecules

The mass spectrum produced from a molecule consists of a number of peaks, with a wide range of masses (m/z values).

The peak with the largest mass is called the molecular ion, or M^+ ion.

Most of these peaks are caused by fragmentation:

$$C_2H_6O^+ \quad \rightarrow \quad C_2H_5^+ \quad + \quad OH$$
M^+ ion $\qquad\qquad$ fragment ion $\qquad$ uncharged fragment (not detected)

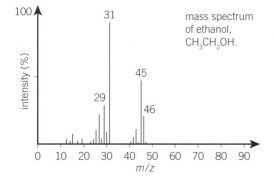

mass spectrum of ethanol, CH_3CH_2OH.

Interpreting the mass spectrum of a molecule

From the mass spectrum you can make various deductions:

- the relative molecular mass of the molecule (from the *m/z* value of the molecular ion)
- the molecular formula of the fragment ions, and of the other fragment that has been lost in order to form the fragment.

Isotope peaks

You may notice on the mass spectrum that there is a tiny peak at 47, just to the right of the peak caused by the M^+ ion.

This is caused by the presence of ^{13}C isotopes in the ion.

About 1% of C atoms are ^{13}C, rather than ^{12}C, so a small fraction of ethanol molecules will contain one ^{13}C atom. The molecular mass of an ethanol molecule with one ^{13}C is 47, not 46.

The peak at 47 is known as the M+1 peak.

Summary questions

1 A sample of an element contains several isotopes. State 3 pieces of information about the isotopes that can be deduced from the mass spectrum of the element. *(3 marks)*

2 A mass spectrum of the organic molecule benzene (C_6H_6) includes a peaks at *m/z* = 78 and a much smaller peak at *m/z* = 79. Explain the origin of these peaks. *(2 marks)*

1 Visible light, ultraviolet radiation, and infrared radiation are all types of electromagnetic radiation.

What is the correct order for the wavelengths of these types of radiation, starting with the shortest:

A visible light, ultraviolet, infrared

B infrared, visible light, ultraviolet

C ultraviolet, infrared, visible light

D ultraviolet, visible light, infrared. *(1 mark)*

2 Which of the following statements about ozone is true?

1 It is broken down into oxygen molecules and oxygen atoms when it absorbs high-frequency ultraviolet radiation.

2 It is formed naturally in the stratosphere.

3 It is formed in the troposphere by reactions involving sunlight and other pollutant gases.

A 1, 2, and 3　　　**B** Only 1 and 2

C Only 2 and 3　　**D** Only 3 *(1 mark)*

3 Sodium ions can be identified using a flame test. Sodium ions produce an intense yellow colour when introduced into a Bunsen flame. The best explanation for this is:

A A photon of yellow light is released when electrons drop from higher to lower energy levels.

B Some frequencies of white light are absorbed when electrons are excited to higher energy levels, leaving only yellow light able to pass through the sodium ions.

C Sodium ions reflect yellow light from the Bunsen flame.

D Sodium ions provide energy to the molecules in the flame, causing their electrons to release photons of yellow light. *(1 mark)*

4 Which of the following statements about the fingerprint region of the spectrum are true?

1 It shows the absorptions of infrared radiation above 1500 cm^{-1}.

2 It is used to identify the bonds in the molecule.

3 Each molecule has a unique pattern in the fingerprint region.

A 1, 2 and 3　　　**B** Only 1 and 2

C Only 2 and 3　　**D** Only 3 *(1 mark)*

5 The molecular ion (M+) in the mass spectrum of a molecule tells you:

A The abundance of the most common isotope

B The relative molecular mass of the most abundant fragment ion

C The relative mass of the most abundant isotope in the molecule

D The relative molecular mass of the unfragmented molecule. *(1 mark)*

6 **a** Ozone, O_3 is a gas present in both the stratosphere and the troposphere:

i The depletion of ozone in the stratosphere has important implications for human health at the Earth's surface.
Explain why. *(2 marks)*

ii Describe the processes that lead to CFC molecules causing ozone depletion in the stratosphere. *(4 marks)*

b Ozone in the troposphere is regarded as a pollutant. One reason for this is that it is involved in a series of reactions that cause photochemical smog. One of these reactions is equation 5.1:

$$NO_2 \rightarrow NO + O$$

    i The reaction in equation 5.1 is caused by the action of sunlight on the NO_2 molecule. Suggest the type of bond breaking that occurs in this reaction. *(1 mark)*

    ii The NO species produced in the reaction is described as a radical. State the meaning of the term radical. *(1 mark)*

    iii The bond enthalpy of the N–O bond in NO_2 is $+305\,kJ\,mol^{-1}$. Calculate the minimum energy, in J, needed to break **one** N–O bond in the NO_2 molecule. *(2 marks)*

    iv Calculate the minimum frequency of light needed to break an N–O bond. *(2 marks)*

    v Suggest how the presence of O atoms enables ozone molecules to form in the troposphere. *(1 mark)*

7 A student oxidises propan-1-ol using sodium dichromate as an oxidising agent. She hopes to form propanal but suspects that the product of the reaction is propanoic acid.

a She obtains an infrared spectrum of the molecule, which shows peaks at the following wavenumbers

2510–$3490\,cm^{-1}$ (very broad)

$1715\,cm^{-1}$

$1415\,cm^{-1}$

$1240\,cm^{-1}$.

    i State what happens to a molecule when it absorbs infrared radiation. *(2 marks)*

    ii Describe two pieces of data that support the student's idea that the product is propanoic acid and not propanal. *(4 marks)*

b The student also obtains a mass spectrum of the product. Peaks are seen at the following *m/z* values: 75, 74, 57, and 45.

    i Explain why peaks are seen at both *m/z* = 74 and *m/z* = 75. *(2 marks)*

    ii Give the formula of the species responsible for the peaks at 57 and 45. *(2 marks)*

Dynamic equilibrium

When the rate of the forward reaction is the same as the rate of the reverse reaction, a system is in dynamic equilibrium, represented by the symbol $\rightleftharpoons$. For example, if you are attempting to dry some wet socks on a radiator, the following process occurs:

$$H_2O(l) \rightarrow H_2O(g)$$

This is a physical change and results in the socks drying as the water evaporates. However, if the socks are in a sealed bag, the water vapour condenses on the inner surface of the bag. In other words, the reverse reaction occurs:

$$H_2O(g) \rightarrow H_2O(l)$$

At first, the rate of the forward reaction is greater than the rate of the reverse reaction, but gradually the rate of the reverse reaction increases. Eventually, both reactions occur at the same rate and a dynamic equilibrium is established. The socks will never fully dry.

$$H_2O(l) \rightleftharpoons H_2O(g)$$

Features of a dynamic equilibrium

- A dynamic equilibrium can only occur in a closed system. Gaseous reactions must take place in a sealed container. Reactions in solution act as a closed system, as long as no reactant can escape.

- Dynamic equilibrium can be established from either direction, starting with the reactants, the products or a mixture.

- Once a dynamic equilibrium has been established, the concentrations of reactants and products remain unchanged. However, the forward reaction and reverse reaction do not stop – their *rates* are equal.

- If most of the reactants become products before the reverse reaction increases sufficiently to establish equilibrium, we say that the position of equilibrium lies to the right. If little of the reactants have changed to products when the reverse reaction becomes equal to the rate of the forward reaction, we say that the position of equilibrium lies to the left.

A chemical equilibrium

When carbon dioxide dissolves in water, the following reversible chemical change occurs:

$$CO_2(aq) + H_2O(l) \rightleftharpoons HCO_3^-(aq) + H^+(aq)$$

This is another example of a dynamic equilibrium. Once equilibrium is established, all four species will be present. The concentrations of each depend on the position of equilibrium under those specific conditions.

Summary questions

1. Explain what is meant by the term 'dynamic equilibrium'. *(1 mark)*

2. Explain why a dynamic equilibrium can be established from either direction, referring to the rates of forward and reverse reaction. *(2 marks)*

3. Explain the effect on the equilibrium between liquid water and water vapour of increasing the temperature of the wet socks in a sealed bag. *(2 marks)*

7.2 Equilibrium constant K_c

Writing an equilibrium constant

Consider the general equilibrium reaction:

$$aA(aq) + bB(aq) \rightleftharpoons cC(aq) + dD(aq)$$

The equilibrium law states that:

$$K_c = \frac{[C]^c[D]^d}{[A]^a[B]^b}$$

This constant, K_c, is the equilibrium constant for the reaction at a specified temperature. The letter K is used to represent all equilibrium constants. When the expression is written in terms of concentrations, we write K_c.

For example, for the equilibrium reaction:

$$N_2(g) + 3H_2(g) \rightleftharpoons 2NH_3(g)$$
$$K_c = \frac{[NH_3]^2}{[N_2][H_2]^3}$$

K_c is a measure of how far a reaction proceeds. If an equilibrium mixture is composed largely of reactants, then the value of K_c is small; if the equilibrium mixture is composed largely of products, then the value of K_c is large.

What affects the value of K_c?

The *only* thing that affects the numerical value of K_c is a change in *temperature*.

	Exothermic reactions	Endothermic reactions
Temperature increases	value of K_c decreases	value of K_c increases
Temperature decreases	value of K_c increases	value of K_c decreases

Summary questions

1. Write expressions for K_c for the following reversible reactions. (*2 marks*)
   a $2SO_2(g) + O_2(g) \rightleftharpoons 2SO_3(g)$
   b $N_2O_4(g) \rightleftharpoons 2NO_2(g)$.

2. Ethanol is produced in industry by the hydration of ethene:
   $$C_2H_4(g) + H_2O(g) \rightleftharpoons C_2H_5OH(g).$$
   The forward reaction is exothermic.
   a Write an expression for K_c for this reaction. (*1 mark*)
   b Explain how increasing the pressure would affect the position of equilibrium and the value of K_c. (*1 mark*)
   c Explain how increasing the temperature would affect the position of equilibrium and the value of K_c. (*1 mark*)
   d State the advantage of using a catalyst, in terms of the equilibrium. (*1 mark*)

Key term

Equilibrium constant: A mathematical expression incorporating the equilibrium concentrations of each substance present.

Revision tip

Remember – products divided by reactants!

Revision tip

a and b are the number of moles of reactants A and B; c and d are the number of moles of products C and D.

Revision tip

The square brackets indicate the concentration in mol dm^{-3}, of whatever is inside the brackets.

Synoptic link

You will find out how to calculate the value of K_c in Topic 7.3, Equilibrium constant and concentrations.

Synoptic link

There is more about factors that affect the value of K_c in Topic 7.4, Le Châtelier's principle.

Calculating an equilibrium constant

Recall the general expression for K_c:

$$K_c = \frac{[C]^c\,[D]^d}{[A]^a\,[B]^b}$$

Calculating the value of K_c is straightforward if the equilibrium concentrations are known.

Worked example: Calculating an equilibrium constant

Calculate the value of K_c at 763 K for the reaction $H_2(g) + I_2(g) \rightleftharpoons 2HI(g)$ given the following data:

$$[H_2(g)] = 1.92\ \text{mol dm}^{-3},\ [I_2(g)] = 3.63\ \text{mol dm}^{-3}\ \text{and}\ [HI(g)] = 17.8\ \text{mol dm}^{-3}.$$

Step 1: Write an expression for K_c.

Step 2: Substitute the values for the equilibrium concentrations of each species:

$$K_c = \frac{[HI]^2}{[H_2][I_2]} = \frac{17.8^2}{1.92\ 3.63} = 45.5.$$

Summary questions

1 Calculate the value of K_c for the Haber process reaction:
   $3H_2(g) + N_2(g) \rightleftharpoons 2NH_3(g)$ at 1 000 K, given $[H_2(g)] = 1.84\ \text{mol dm}^{-3}$, $[N_2(g)] = 1.36\ \text{mol dm}^{-3}$, and $[NH_3(g)] = 0.142\ \text{mol dm}^{-3}$. (*2 marks*)

2 Calculate the concentration of ethanol at equilibrium in the esterification reaction:
   $C_2H_5OH(l) + CH_3COOH(l) \rightleftharpoons CH_3COOC_2H_5(l) + H_2O(l)$
   given that $K_c = 4.10$, $[CH_3COOH] = 0.80\ \text{mol dm}^{-3}$, and $[CH_3COOC_2H_5] = [H_2O] = 3.0\ \text{mol dm}^{-3}$. (*2 marks*)

Worked example: Composition of equilibrium mixtures

You can also calculate the composition of equilibrium mixtures:

$$CH_3COOH(l) + C_2H_5OH(l) \rightleftharpoons CH_3COOC_2H_5(l) + H_2O(l).$$

K_c for the above esterification reaction has a value of 4.10 at 25 °C.

Calculate the equilibrium concentration of ethyl ethanoate, given that $[CH_3COOH(l)] = 0.255\ \text{mol dm}^{-3}$, $[C_2H_5OH(l)] = 0.245\ \text{mol dm}^{-3}$ and $[H_2O(l)] = 0.437\ \text{mol dm}^{-3}$.

Step 1: Write an expression for K_c for the reaction.

Step 2: Rearrange the K_c expression to make the unknown concentration the subject of the equation.

Step 3: Substitute the values of the known concentrations and of K_c.

$$K_c = \frac{[CH_3COOC_2H_5]\,[H_2O]}{[CH_3COOH]\,[C_2H_5OH]}$$

Rearranging the expression:

$$[CH_3COOC_2H_5] = \frac{K_c \times [CH_3COOH] \times [C_2H_5OH]}{[H_2O]}$$

Substituting the values into the rearranged expression:

$$[CH_3COOC_2H_5] = \frac{4.10 \times 0.255\ \text{mol dm}^{-3} \times 0.245\ \text{mol dm}^{-3}}{0.437\ \text{mol dm}^{-3}}$$

$$= 0.586\ \text{mol dm}^{-3}.$$

7.4 Le Châtelier's principle

Specification reference: ES(q)

Le Châtelier's principle

The position of the equilibrium can be altered by changing the concentration of solutions, the pressure of gases, or the temperature.

Le Châtelier's principle states that if a system is at equilibrium, and a change is made in any of the conditions, then the system responds to counteract the change as much as possible.

Concentration

Concentration change	Equilibrium shift
increasing reactant(s)	to the right (decreases reactants)
increasing product(s)	to the left (decreases products)
decreasing reactant(s)	to the left (increases reactants)
decreasing product(s)	to the right (increases products)

Pressure

Pressure change	Equilibrium shift
Increasing	to the side with fewer gas molecules – in the example below, to the right. $CO(g) + 2H_2(g) \rightleftharpoons CH_3OH(g)$ 3 molecules 1 molecule
Decreasing	to the side with more gas molecules – in the example below, to the right $CH_4(g) + H_2O(g) \rightleftharpoons CO(g) + 3H_2(g)$ 2 molecules 4 molecules

Temperature

Temperature change	Equilibrium shift
Increase	position of equilibrium shifts in the direction of the endothermic reaction
Decrease	position of equilibrium shifts in the direction of the exothermic reaction

> **Revision tip**
>
> A catalyst does not change the position of equilibrium, just the rate at which the equilibrium is established.

> **Common misconception: Changing pressue or concentration**
>
> Changing the pressure or concentration does not affect the magnitude of K_c. Only temperature affects the magnitude of K_c.

> **Common misconception: Position of equilibrium**
>
> Remember to say that the *position of equilibrium* shifts.

Summary questions

1 $C_2H_4(g) + H_2O(g) \rightleftharpoons C_2H_5OH(g)$; $\Delta H = -46 kJ\,mol^{-1}$.
   Predict, using le Châtelier's principle, the changes in the following that would move the position of equilibrium to the right:
   a concentration   (1 mark)
   b temperature   (1 mark)
   c pressure.   (1 mark)

2 $Fe^{3+}(aq) + SCN^-(aq) \rightleftharpoons [FeSCN]^{2+}(aq)$
   pale yellow   colourless   blood red
   For this reaction, what would you see if you added:
   a more $Fe^{3+}(aq)$ (1 mark)
   b more $[FeSCN]^{2+}(aq)$?   (1 mark)

3 Study the equilibrium reactions below. Predict the effect on the position of equilibrium of increasing the temperature and of increasing the pressure.
   a $N_2(g) + 3H_2(g) \rightleftharpoons 2NH_3(g)$; $\Delta H = -92 kJ\,mol^{-1}$ (2 marks)
   b $N_2(g) + O_2(g) \rightleftharpoons 2NO(g)$; $\Delta H = +90 kJ\,mol^{-1}$ (2 marks)

1 Which statement gives the best description of the term 'dynamic equilibrium'?

   A A reaction which goes in both directions.

   B A reversible reaction in which the concentrations of the reactants and the products are equal.

   C A reversible reaction in a closed system in which the rate of the forward reaction is equal to the rate of the reverse reaction.

   D A reaction which is static and has no change in the concentration of reactants and products. *(1 mark)*

2 Which expression shows the equilibrium concentration of PF_3 in the reaction:

$$PF_5 \rightleftharpoons PF_3 + F_2?$$

   A $[PF_3] = \dfrac{[PF_3][F_2]}{[PF_5]}$, **B** $[PF_3] = \dfrac{K_c[F_2]}{[PF_5]}$, **C** $[PF_3] = \dfrac{[F_2]}{K_c[PF_5]}$, **D** $[PF_3] = \dfrac{K_c[PF_5]}{[F_2]}$

(1 mark)

3 Ammonia reacts with water as follows:

$$NH_3 + H_2O \rightleftharpoons NH_4^+ + OH^-$$

   a Suggest the likely pH of a solution of ammonia in water. *(1 mark)*

   b Use le Châtelier's principle to predict and explain how the position of equilibrium changes when an aqueous solution of an acid is added to the equilibrium mixture. *(2 marks)*

4 It is possible to produce liquid fuels from methane. This is only really feasible for countries that have large reserves of natural gas. The first stage is the production of methanol in a two-step process.

   Step 1 $CH_4(g) + H_2O(g) \rightleftharpoons CO(g) + 3H_2(g)$ $\Delta H = +205\,kJ\,mol^{-1}$

   Step 2 $CO(g) + 2H_2(g) \rightleftharpoons CH_3OH(g)$ $\Delta H = -125\,kJ\,mol^{-1}$

Step	Temperature (K)	Pressure (atm)	Catalyst used
1	1050	5	nickel
2	540	90	copper

   a i Describe and explain how temperature, pressure, and the presence of a catalyst affect the position of equilibrium for the reactions in **step 1** and **step 2**. *(6 marks)*

   ii Suggest and explain why the temperature used for the reaction shown in **step 2** is lower than for the reaction shown in **step 1**. *(3 marks)*

   iii Suggest and explain why the pressure used for the reaction shown in **step 2** is higher than for **step 1**. *(3 marks)*

   b Most of the methanol produced at the end of **step 2** is not used as a fuel, but is converted into a mixture of hydrocarbons in another two-step process.

   Step 3 $2CH_3OH(g) \rightleftharpoons CH_3OCH_3(g) + H_2O(g)$

   Step 4 $4CH_3OCH_3(g) \rightleftharpoons C_8H_{16}(g) + 4H_2O(g)$

   i Write an expression for the equilibrium constant, K_c, for the reaction shown in **step 3**. *(2 marks)*

   ii Use the data given below to calculate a value for K_c at 600 K for the reaction shown in **step 3**. *(2 marks)*

     $[CH_3OH(g)] = 0.050\,mol\,dm^{-3}$; $[CH_3OCH_3(g)] = 0.20\,mol\,dm^{-3}$; $[H_2O(g)] = 0.20\,mo\,dm^{-3}$

8.1 Acids, bases, alkalis, and neutralisation

Specification reference: EL(t), EL(c (i))

Acids, bases, and alkalis

An acid is a compound that dissociates in water to produce hydrogen ions. For example, hydrochloric acid dissociates as follows:

$$HCl(aq) \rightarrow H^+(aq) + Cl^-(aq)$$

Bases are compounds that react with an acid to produce water and a salt. An alkali is a base that dissolves in water to produce hydroxide, OH^- ions. For example, sodium hydroxide is an alkali:

$$NaOH(aq) \rightarrow Na^+(aq) + OH^-(aq)$$

The properties of acids are a result of the transfer of H^+ ions to bases. Acids are sometimes referred to as proton donors and bases as proton acceptors.

Neutralisation reactions can be summarised using ionic equations. When hydrochloric acid reacts with sodium hydroxide the ions involved, and the products, are:

$$H^+(aq) + Cl^-(aq) + Na^+(aq) + OH^-(aq) \rightarrow H_2O(l) + Na^+(aq) + Cl^-(aq)$$

When the spectator ions have been removed, the ionic equation is:

$$H^+(aq) + OH^-(aq) \rightarrow H_2O(l)$$

Calculations of concentrations

Concentrations can be measured in grams per cubic decimetre ($g\,dm^{-3}$), but it is more usual to use moles per cubic decimetre ($mol\,dm^{-3}$).

Recall the formulae for concentration calculations:

$$C = \frac{n}{V}$$
$$n = c \times V$$
$$V = \frac{n}{c}$$

c is the concentration in $mol\,dm^{-3}$

V is the volume in dm^3

n is the amount in moles.

Any of the three quantities (concentration, amount, or volume) can be calculated by using the correct expression and the other two known values.

Common misconception: cm³ to dm³

Always remember to convert volumes to dm^3. Divide cm^3 by 1000 to get the volume in dm^3.

Key term

Acid: Produces H^+ ions in solution.

Key term

Base: A compound that reacts with an acid to produce water and a salt.

Revision tip

All acid/alkali neutralisation reactions have this ionic equation, whatever acid and alkali are involved.

Synoptic link

You can find more about calculating concentration in Topic 1.4, Concentrations of solutions.

Worked example: Concentration calculations

Concentration	Amount	Volume
$c = \dfrac{n}{V}$	$n = c \times V$	$V = \dfrac{n}{c}$
What is the concentration of a solution of 0.5 mole of NaOH in 100 cm^3?	How many moles are in 20 cm^3 of 0.1 mol dm^{-3} NaOH?	What volume of 0.2 mol dm^{-3} NaOH contains 0.1 mol?
$c = \dfrac{0.5}{0.1}$	$n = 0.1 \times 0.02$	$V = \dfrac{0.1}{0.2}$
$= 5$ mol dm^{-3}	$= 0.002$ mol	$= 0.5$ dm^3

Synoptic link

Details of how to carry out a titration calculation are given in Topic 1.4, Concentrations of solutions.

Using concentrations in calculations

The concentration of a solution can be determined using a titration. The *end-point* of the reaction is often detected by the use of an indicator that changes colour.

Making salts

Salts are produced by the reaction of bases or alkalis with acids. For example:

sodium carbonate + hydrochloric acid → sodium chloride + water + carbon dioxide

potassium hydroxide + nitric acid → potassium nitrate + water

The type of salt produced depends on the alkali and acid used. Hydrochloric acid produces chlorides, nitric acid produces nitrates, and sulfuric acid produces sulfates.

Salts produced in this way are generally soluble. The solid salt can be produced by evaporating the solution.

Some salts are insoluble and can be produced by precipitation reactions. For example, silver iodide can be produced by reacting silver nitrate and potassium iodide.

Summary questions

1 Convert the following volumes into dm^3:
   a 50 cm^3
   b 2500 cm^3. (2 marks)
2 Calculate the concentration (in mol dm^{-3}) of the following solutions:
   a 40 g of NaOH in 500 cm^3 of solution (1 mark)
   b 10.6 g of Na$_2$CO$_3$ in 2000 cm^3 of solution. (1 mark)

3 Calculate the mass of solute needed to make up the following solutions:
   a 100 cm^3 of a 0.5 mol dm^{-3} solution of MgCl$_2$ (1 mark)
   b 2 dm^3 of a 0.02 mol dm^{-3} solution of KMnO$_4$. (1 mark)
4 A volume of 20 cm^3 sulfuric acid was used in a titration, delivered from a burette. The concentration of the solution was 0.100 mol dm^{-3}.
   a Calculate the volume of acid used in dm^3. (1 mark)
   b Calculate the number of moles of acid used. (1 mark)
   c Deduce how many moles of NaOH this acid would neutralise.
      $H_2SO_4(aq) + 2NaOH(aq) \rightarrow Na_2SO_4(aq) + 2H_2O(l)$ (1 mark)
   d If this number of moles of NaOH were in a 25.0 cm^3 sample, calculate the concentration of the sodium hydroxide solution, in mol dm^{-3}. (1 mark)

Chapter 8 Practice questions

1 Which of the following is **not** an acid?

   A HNO_3 **B** $Ca(OH)_2$

   C CH_3COOH **D** H_3PO_4 *(1 mark)*

2 Which is the general equation for neutralisation?

   A $H^+ + OH^- \rightarrow H_2O$

   B $HCl + NaOH \rightarrow H_2O + NaCl$

   C $H_2O \rightarrow H^+ + OH^-$

   D $HCl \rightarrow H^+ + Cl^-$ *(1 mark)*

3 What is the concentration of a solution containing 4 g of NaOH in 500 cm^3 of water?

   A $2\,g\,dm^{-3}$ **B** $4\,g\,dm^{-3}$

   C $8\,g\,dm^{-3}$ **D** $2000\,g\,dm^{-3}$ *(1 mark)*

4 What is the concentration of a solution containing 23.8 g of KBr in 100 cm^3 of water?

   A $0.002\,mol\,dm^{-3}$ **B** $0.2\,mol\,dm^{-3}$

   C $2\,mol\,dm^{-3}$ **D** $20\,mol\,dm^{-3}$ *(1 mark)*

5 25.0 cm^3 of 0.01 $mol\,dm^{-3}$ HCl required 19.75 cm^3 of a solution of NaOH for complete neutralisation.

   a Calculate the amount in moles of hydrochloric acid used.   *(1 mark)*

   b Deduce the amount in moles of sodium hydroxide used.   *(1 mark)*

   c Calculate the concentration of the sodium hydroxide in $mol\,dm^{-3}$.   *(1 mark)*

6 25.0 cm^3 of 0.2 $mol\,dm^{-3}$ H_2SO_4 required 28.15 cm^3 of a solution of KOH for complete neutralisation.

$$H_2SO_4 + 2KOH \rightarrow K_2SO_4 + 2H_2O$$

   a Calculate the amount in moles of sulfuric acid used.   *(1 mark)*

   b Deduce the amount in moles of potassium hydroxide used.   *(1 mark)*

   c Calculate the concentration of the potassium hydroxide in $mol\,dm^{-3}$.   *(1 mark)*

7 Household bleach solutions contain the chlorate(I) ion, ClO^-, as the active ingredient. The chlorate(I) ion content can be determined by titration. An analyst carries out a titration as follows:

A 5.00 cm^3 sample of bleach is reacted with an excess of acidified potassium iodide. The solution goes deep brown.

$$ClO^- + 2I^- + 2H^+ \rightarrow Cl^- + I_2 + H_2O$$

This mixture is then titrated with 0.100 $mol\,dm^{-3}$ sodium thiosulfate solution, $Na_2S_2O_3(aq)$. 25.40 cm^3 of thiosulfate solution are needed to obtain the colourless end-point.

$$2S_2O_3^{2-} + I_2 \rightarrow S_4O_6^{2-} + 2I^-$$

   a **i** Calculate the number of moles of thiosulfate, $S_2O_3^{2-}$, used in the titration.   *(2 marks)*

      ii How many moles of iodine have reacted with the thiosulfate?   *(1 mark)*

      iii How many moles of chlorate(I) are in the sample of bleach?   *(1 mark)*

      iv What is the concentration of chlorate(I) in the bleach, in $mol\,dm^{-3}$?   *(3 marks)*

   b Calculate the concentration of the chlorate(I) ion, in $g\,dm^{-3}$, in the sample of bleach.   *(3 marks)*

Key terms

Redox reaction: A reaction involving oxidation and reduction simultaneously.

Oxidation: Oxidation is the loss of electrons.

Reduction: Reduction is the gain of electrons.

Redox reactions

When an oxidation reaction and a reduction reaction occur simultaneously, this is called a **redox reaction**.

Oxidation and reduction can be defined in two different ways:

- **o**xidation **i**s the **l**oss of electrons
- **r**eduction **i**s the **g**ain of electrons

or

- an element is oxidised when its oxidation state is increased (becomes more positive)
- an element is reduced when its oxidation state is decreased (becomes more negative).

Electron transfer and half-equations

Sodium reacts with chlorine as follows:

$$2Na + Cl_2 \rightarrow 2NaCl$$

This can be written as two separate half-equations. From these you can decide which is the oxidation reaction and which is the reduction reaction:

$$2Na \rightarrow 2Na^+ + 2e^-$$ This is oxidation (electron loss)

$$Cl_2 + 2e^- \rightarrow 2Cl^-$$ This is reduction (electron gain)

Assigning oxidation states

Oxidation states help determine which species are oxidised and which are reduced, even when no ions are present.

You should learn the following rules:

- The atoms in elements are always in an oxidation state of zero.
- In compounds or ions, oxidation states are assigned to each atom or ion.

Since compounds have no overall charge, the oxidation states of all the constituents must add up to zero. In ions, the sum of the oxidation states is equal to the charge on the ion.

Some atoms rarely change their oxidation states in reactions. These can be used to help to assign oxidation states to other species. Examples are F is -1, O is -2 (except in O_2^{2-} and OF_2), H is $+1$, Cl is usually -1 (except when combined with O or F).

Revision tip
Remember OIL RIG: Oxidation Is Loss; Reduction Is Gain.

🖩 Worked example: Oxidation states

Assign the oxidation states for each element in the following compounds:

a CO_2

Step 1: O is -2; there are two Os so the total contribution of O to the oxidation state is $2 \times (-2) = -4$.

Step 2: To make the total of the oxidation states add up to zero, C must be $+4$.

▶

b CH_4

Step 1: H is +1; there are four hydrogens, so the total contribution of H to the oxidation state is $4 \times (+1) = +4$.

Step 2: To make the total of the oxidation states add up to zero, C is −4.

c VO^{2+}

Step 1: O is −2.

Step 2: The charge on the ion is +2, so V must have an oxidation state of +4, because $+4 + (-2) = +2$.

Using oxidation states

The displacement reaction below can be used to extract iodine from iodide minerals present in seawater. A more reactive halogen, like Cl_2, is passed into a solution of iodide ions, which are less reactive:

$$Cl_2(aq) + 2I^-(aq) \rightarrow 2Cl^-(aq) + I_2(aq)$$

oxidation states: 0   −1   −1   0

- The oxidation state for chlorine decreases (from 0 to −1), so chlorine is reduced.
- The oxidation state for iodine increases (from −1 to 0), so iodine is oxidised.

This reaction can also be represented by two half-equations:

$$Cl_2(aq) + 2e^- \rightarrow 2Cl^-(aq) - \text{this is reduction (electron gain)}$$

$$2I^-(aq) \rightarrow I_2(aq) + 2e^- - \text{this is oxidation (electron loss)}$$

Common misconception: Oxidation

Remember the correct word is oxidation, not 'oxidisation'!

The chlorine molecules are acting as the oxidising agent, as they cause another species to be oxidised and in doing so are reduced themselves. The iodide ions are reducing agents as they cause another species to be reduced and in doing so are oxidised themselves.

Using oxidation states to balance redox equations

To balance redox equations, identify the change in oxidation state (if any) of each element present. Then balance the equation so the number of electrons lost is equal to the electrons gained.

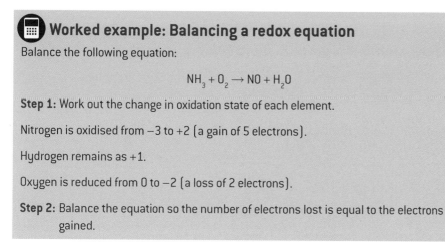

🖩 Worked example: Balancing a redox equation

Balance the following equation:

$$NH_3 + O_2 \rightarrow NO + H_2O$$

Step 1: Work out the change in oxidation state of each element.

Nitrogen is oxidised from −3 to +2 (a gain of 5 electrons).

Hydrogen remains as +1.

Oxygen is reduced from 0 to −2 (a loss of 2 electrons).

Step 2: Balance the equation so the number of electrons lost is equal to the electrons gained.

Revision tip

You must always write down the sign of an oxidation state (+ or −), or you will lose marks. The sign is written *in front of* the oxidation number.

Key terms

Half-equation: An ionic equation showing the transfer of electrons from one type of atom. Two half-equations combine together to make the chemical equation for the reaction.

Oxidising agent: A species that causes another substance to become oxidised. Oxidising agents become reduced as they take electrons from the other substance.

Reducing agent: A species that causes another substance to become reduced. Reducing agents become oxidised as they provide electrons for the other substance.

Two nitrogen atoms can gain a total of 10 electrons from 5 oxygen atoms. Each oxygen atom loses two electrons.

Two ammonia molecules are therefore required. This will produce two NO and three H_2O molecules.

The balanced equation is:

$$2NH_3 + 2\tfrac{1}{2} O_2 \rightarrow 2NO + 3H_2O$$

This can also be written as:

$$4NH_3 + 5 O_2 \rightarrow 4NO + 6H_2O.$$

Revision tip

The oxidation state for cations is written in Roman numerals, immediately *after* the element it refers to.

Oxidation states in names

Some compounds contain elements that can exist in more than one oxidation state. When this occurs, the systematic name for the compound includes the oxidation state of the element. For example, FeO is called iron(II) oxide and Fe_2O_3 is called iron(III) oxide.

Oxoanions are negative ions that contain oxygen and another element. Their names end with the letters '-ate', for example, chromate. The names of oxoanions should also include an oxidation state. For example:

Ion	Name	Oxidation state of N or S
NO_2^-	nitrate(III)	+3
NO_3^-	nitrate(V)	+5
SO_3^{2-}	sulfate(IV)	+4
SO_4^{2-}	sulfate(VI)	+6

Summary questions

1 Write down the oxidation states of the elements in the following.

(6 marks)

   a KBr   **b** H_2O **c** CO
   d PO_4^{3-} **e** MnO_2 **f** $Cr_2O_7^{2-}$

2 Write down the formulae of the following. *(4 marks)*
   a copper(II) chloride   **b** copper(I) oxide
   c lead(IV) chloride   **d** manganate(VII) ion

3 Write two half-equations for the following reaction, and identify which equation is the oxidation reaction and which is the reduction reaction. *(4 marks)*

$$2Ca + O_2 \rightarrow 2CaO$$

4 Write a balanced equation for the following reaction by identifying the changes in oxidation states. *(2 marks)*

$$Br^- + H^+ + H_2SO_4 \rightarrow Br_2 + SO_2 + H_2O.$$

9.2 Electrolysis as redox

Specification reference: ES(c)

Electrolysis of molten salts

Molten salts conduct electricity, because they contain ions which are free to move. The metal ions, which are positive, are attracted to the cathode where they gain electrons and turn into the metal:

$$Pb^{2+}(l) + 2e^- \rightarrow Pb(l)$$

The non-metal ions, which are negative, are attracted to the anode where they lose electrons and turn into the element.

$$2Br^-(l) \rightarrow Br_2(g) + 2e^-$$

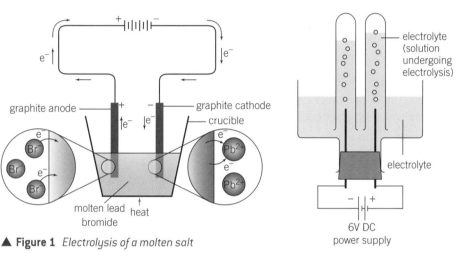

▲ **Figure 1** *Electrolysis of a molten salt*

▲ **Figure 2** *Electrolysis of a solution of a salt*

Key terms

Cations: Positive ions are known as *cations*, as they are attracted to the *cathode*.

Anions: Negative ions are known as *anions*, as they are attracted to the *anode*.

Reduction occurs at the cathode, because electrons are gained. Oxidation occurs at the anode.

Electrolyte	Cathode process	Anode process
Molten aluminium oxide	$Al^{3+}(l) + 3e^- \rightarrow Al(l)$	$2O^{2-}(l) \rightarrow O_2(g) + 4e^-$
Molten zinc iodide	$Zn^{2+}(l) + 2e^- \rightarrow Zn(l)$	$2I^-(l) \rightarrow I_2(g) + 2e^-$
Molten magnesium chloride	$Mg^{2+}(l) + 2e^- \rightarrow Mg(l)$	$2Cl^-(l) \rightarrow Cl_2(g) + 2e^-$

Electrolysis of solutions

When an electric current is passed through a solution of a salt, such as sodium chloride, positive ions are attracted to the cathode and negative ions are attracted to the anode. However, when predicting the products, you must take into account the presence of water, which can compete with the ions from the salt at the electrodes. If the solution contains a more-reactive metal such as sodium or potassium, hydrogen gas is produced. Less-reactive metals such as zinc or copper are deposited on the cathode. If the solution contains a halide ion, the corresponding halogen is produced at the anode. If the solution contains an anion such as sulfate or nitrate, oxygen gas is produced.

Using metal electrodes with metal ions in solution

If the anode is made of the same metal as the ions in solution, oxidation occurs and the anode gradually dissolves. For example:

$$Cu(s) \rightarrow Cu^{2+}(aq) + 2e^-$$

The ions produced are then deposited at the cathode.

Summary questions

1 State the products of the electrolysis of these molten salts:
   a potassium chloride *(1 mark)*
   b calcium oxide *(1 mark)*
   c magnesium iodide. *(1 mark)*

2 State the products of the electrolysis of these aqueous solutions with graphite electrodes:
   a potassium chloride solution *(1 mark)*
   b sodium sulfate solution *(1 mark)*
   c copper nitrate solution. *(1 mark)*

3 Write half-equations for the electrolysis of:
   a molten aluminium chloride *(1 mark)*
   b aqueous lead nitrate solution *(1 mark)*
   c aqueous sodium chloride solution. *(1 mark)*

Electrolysis of molten compounds

Solid ionic compounds do not conduct electricity since ions in a giant ionic lattice are not free to move. On melting, the ions are free to move and the liquid can carry a current. For electrolysis to occur, an electric current is passed through the electrolyte, using a power supply and electrodes (anode and cathode) which are commonly made of graphite.

Electrolysis of aqueous solutions

For electrolysis to occur in an aqueous solution an electrical circuit needs to be set up using a dc power supply, as in the electrolysis of a molten salt. The dc power supply may be a powerpack or batteries. Graphite is commonly used for the electrodes as it is cheap and inert. Platinum may be used instead, but it is much more expensive.

If the products of electrolysis are gases, an alternative arrangement can be used to collect the gases (see Figure 1). The test tubes are filled with water and the gaseous products displace the water and fill the test tubes.

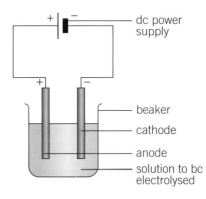

▲ **Figure 1** *Electrolysis of aqueous solutions*

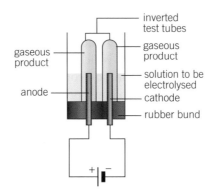

▲ **Figure 2** *Apparatus to collect gaseous products in electrolysis*

1 Which statement is correct about the reaction below?

$Mg + Cl_2 \rightarrow MgCl_2$

  A Magnesium is oxidised

  B Magnesium gains electrons

  C Chlorine is oxidised

  D Chlorine's oxidation state does not change.   *(1 mark)*

2 Which species contains phosphorus in oxidation state +5?

  A P_5 **B** PF_6^-

  C H_3PO_3 **D** PH_3 *(1 mark)*

3 What is the oxidiation state of sulfur in H_2SO_4?

  A +2   **B** +4

  C +6   **D** +8   *(1 mark)*

4 What is the name of $NaNO_2$?

  A sodium nitrate(I)   **B** sodium nitrate(II)

  C sodium nitrate(III)   **D** sodium nitrate(V)   *(1 mark)*

5 What are the products of electrolysis of aqueous potassium nitrate?

  A potassium at the cathode, nitrogen at the anode

  B nitrogen at the anode, potassium at the cathode

  C oxygen at the cathode, nitrogen at the anode

  D hydrogen at the cathode, oxygen at the anode   *(1 mark)*

6 Which substances produce copper at the cathode and oxygen at the anode during electrolysis?

    1 Aqueous copper sulfate

    2 Aqueous copper nitrate

    3 Molten copper chloride

  A 1 only   **B** 1 and 2 only

  C 2 and 3 only   **D** 1, 2, and 3   *(1 mark)*

7 Balance the equation below using oxidation numbers:

$$I^- + H^+ + MnO_2 \rightarrow I_2 + H_2O + Mn^{2+}$$

                                  (2 marks)

8 Identify which species are oxidised and which are reduced in the equation below:

$$2 I_2 + N_2H_4 \rightarrow 4 HI + N_2$$

                                  (2 marks)

9 The water in the Dead Sea is particularly rich in bromide ions, and an important chemical industry has grown up in Israel to exploit this natural resource.

The process involves oxidising bromide ions into bromine molecules.

  a Describe the colour change you would expect to see when Br^- is converted to Br_2.   *(2 marks)*

  b Explain why the process in which bromide ions, Br^- are changed into bromine, Br_2 is referred to as *oxidation*.   *(1 mark)*

  c Chlorine is used to oxidise the bromide ions. Chlorine is manufactured on site by the electrolysis of brine. Write a half-equation for the production of chlorine from brine.   *(1 mark)*

  d Write an equation for the reaction of chlorine with bromide ions.   *(1 mark)*

  e Explain why chlorine is described as an oxidising agent, referring to changes in the oxidation state of chlorine.   *(2 marks)*

Revision tip

Increasing the concentration, pressure, temperature, surface area, or intensity of radiation will increase the rate of reaction.

Common misconception: Frequency of collisions

It is important to say that there are more frequent collisions, not just more collisions.

Key term

Collision theory: The theory that explains how the rate of chemical reactions is governed by the frequency of collisions between particles.

Revision tip

The more frequent collisions, the faster the rate of reaction.

Rates of reaction

Rates of reaction can be affected by a number of factors:

- concentration
- pressure
- a catalyst
- temperature
- surface area
- particle size
- intensity of radiation.

Collision theory

Reactions occur when particles of reactants collide with a certain *minimum* kinetic energy:

- At higher concentrations and higher pressures, the particles are in closer proximity to each other, encouraging more frequent collisions.
- At higher temperatures, a much higher proportion of colliding particles have sufficient energy to react and more collisions have a greater energy than the activation enthalpy.
- When a solid is more finely divided, there is a larger surface area on which the reactions can take place, so the greater the frequency of successful collisions.
- Heterogeneous catalysts provide a surface where reacting particles may break and make bonds.

All of the above serve to increase the rate of a chemical reaction.

Measuring rate of reaction

To measure the reaction rate you need to measure how quickly a reactant is used up, or how quickly a product is made. This can be done by:

- Measuring volume of gas produced. This can be done using a gas syringe, or by displacement of water. The more gas produced per unit time, the faster the reaction.
- Measuring mass changes. Reactions which give off a gas involve mass changes. The bigger the change in mass per unit time, the faster the reaction.
- Colorimetry. Measuring the change of intensity as a coloured chemical is used up or produced. For example, a reaction involving the production of bromine will involve a gradually darkening brown colouration. The rate of change of intensity of the brown colouration can be measured using a colorimeter.
- pH changes. If an acid or alkali is used up or produced, a pH meter can monitor the rate of change of pH. The faster the change, the faster the reaction.

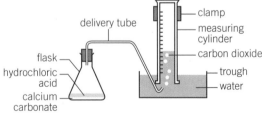

▲ **Figure 1** *Measuring volume changes*

10.2 The effect of temperature on rate

Specification reference: OZ[e], OZ [f]

Activation enthalpy

Rates of reaction do not just depend on how frequently particles collide, but also on how much energy they have when a collision takes place.

Collision theory states that reactions occur when molecules collide with a certain *minimum* kinetic energy. This minimum kinetic energy is called the **activation enthalpy**. The energy needed to overcome the energy barrier is called the activation energy barrier.

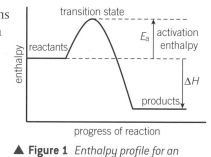

▲ **Figure 1** *Enthalpy profile for an exothermic reaction*

Molecular energies

As the temperature increases, the rate of a chemical reaction also increases. This is because of the distribution of energies among the reacting particles – this distribution is called the **Maxwell–Boltzmann distribution**.

There needs to be enough molecules with sufficient energy for a reaction to take place. The molecules need to have a combined kinetic energy higher than the activation enthalpy.

Reactions go faster at higher temperatures, because a larger proportion of the colliding particles have the minimum activation enthalpy needed to react.

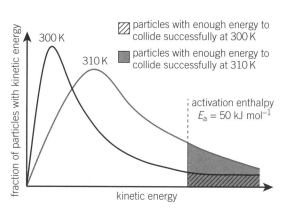

▲ **Figure 2** *Maxwell-Boltzmann distribution for a reaction at two different temperatures*

Referring to the diagram above:

- the peak of the number of collisions at 300 K is at a lower kinetic energy value than the peak at 310 K (i.e. the most probable kinetic energy for a particle is lower at lower temperatures)
- at the kinetic energy value of 50 kJ mol^{-1}, the number of collisions at 310 K is almost twice as many as at 300 K
- for reactions with an activation enthalpy around 50 kJ mol^{-1}, when the temperature rises by 10 °C (10 K), the rate of reaction approximately doubles.

Summary questions

1. Explain what E_a represents. *(1 mark)*

2. Sketch a graph showing a typical distribution of energy among the molecules of a reaction mixture. Shade the area of the graph representing the number of collisions with sufficient energy to lead to a reaction. *(2 marks)*

3. Suggest a reason for the following observations:
   a. N_2 and O_2 do not react at room temperature, but in a car engine the temperature is high enough for them to form NO. *(1 mark)*
   b. The reaction between NO and O_2 to make NO_2 occurs easily at room temperature. *(1 mark)*

Catalysts and activation energy

Catalysts speed up the rate of a chemical reaction without getting used up.

In heterogeneous catalysis, the reactants and catalysts are in different physical states. The effectiveness of heterogeneous catalysts can be reduced by catalyst poisons. In homogeneous catalysis, the reactants and catalysts are in the same physical state. Homogeneous catalysts work by forming an intermediate compound with the reactants. In the first step, an *intermediate* is formed in a reaction with lower activation enthalpy. In the second step, this intermediate breaks down to give a product and reform the catalyst.

For a chemical reaction to proceed, a pair of reacting molecules must collide, with a combined energy greater than the activation enthalpy for the reaction, in order to make a successful collision. With a catalyst, successful collisions can take place at a lower energy.

The reaction profiles for a catalysed reaction and an uncatalysed reaction are shown in Figure 1.

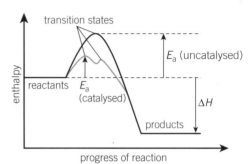

▲ **Figure 1** *The effect of a catalyst on the enthalpy profile of a reaction*

Catalysts work by providing an alternative reaction pathway for the breaking and making of bonds. This alternative path has a lower activation enthalpy than the uncatalysed pathway.

Maxwell-Boltzmann distribution

You can see in the Maxwell-Boltzmann distribution shown in Figure 2 that E_a for the catalysed reaction is lower than for the uncatalysed reaction. Therefore more particles have energy greater than the activation enthalpy and consequently the rate of reaction is greater.

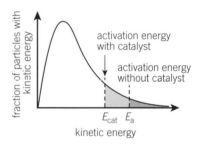

▲ **Figure 2** *Maxwell-Boltzmann distribution showing activation enthalpies for a catalysed and an uncatalysed reaction*

Summary questions

1 Hydrogen peroxide slowly decomposes into water and oxygen.
   a Explain why the activation enthalpy is lower for the decomposition of hydrogen peroxide solution in the presence of manganese(IV) oxide. *(1 mark)*
   b A solution of the enzyme catalase can also be used to decompose hydrogen peroxide solution. Explain what *type* of catalyst is
      i manganese(IV) oxide *(1 mark)*
      ii catalase. *(1 mark)*

2 Look again at the reaction profile in Figure 1. Explain why there is a trough in the pathway for the catalysed reaction. *(2 marks)*

3 Suggest whether the value of the enthalpy change, ΔH, is affected by the use of a catalyst. *(2 marks)*

Heterogeneous catalysts

Heterogeneous catalysts are in a different physical state to the reactants and products. For example, the hydrogenation of ethene to produce ethane uses a nickel catalyst. Nickel is a solid, whereas hydrogen, ethene, and ethane are gases.

Heterogeneous catalysts provide a surface on which a reaction may take place, thus lowering the energy needed for a successful collision – this lowers the activation energy barrier.

This happens in four stages:

1　The reactant molecules are adsorbed onto the surface of the catalyst. This weakens the bonds in the reactant.

2　Bonds in the reactant break.

3　New bonds form. This creates the products.

4　The products are released from the catalyst surface and diffuse away.

Catalyst poisons

Catalysts may be poisoned if a molecule becomes adsorbed more strongly than the reactant molecules. This means that fewer reactant molecules can be adsorbed, and the catalyst becomes less efficient.

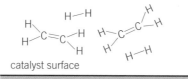

catalyst surface

Reactants get adsorbed onto catalyst surface. Bonds are weakened.

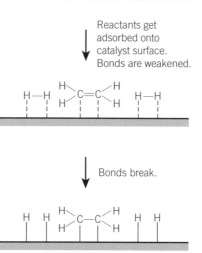

Bonds break.

New bond forms.

Second bond forms, and product diffuses away from catalyst surface, leaving it free to absorb fresh reactants.

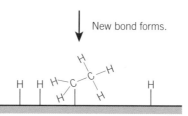

▶ **Figure 1** *A model of heterogeneous catalysis*

Common misconception: Adsorption

'Adsorbed' means to be attached to a solid surface. Do not confuse 'adsorb' and 'absorb'.

Revision tip

Heterogeneous catalysts must have a large surface area in order to adsorb reactants.

They are used in a finely divided form, sometimes supported on a porous material.

Key term

Catalyst poison: A substance which binds irreversibly to the catalyst surface and stops the catalyst functioning properly.

Revision tip

An example of a catalyst poison is lead, which poisons the metals platinum and rhodium found in catalytic converters in cars.

Summary questions

1　Explain which of the following are examples of heterogeneous catalysis. 　(4 marks)
　　a　the use of iron oxides in the reaction of nitrogen and hydrogen to make ammonia.
　　b　breakdown of proteins into amino acids by the protease enzymes.
　　c　the use of vanadium oxides in the oxidation of sulfur dioxide to form sulfur trioxide in the production of sulfuric acid.
　　d　the use of sulfuric acid in the formation of esters from carboxylic acids and alcohols.

2　Describe the stages of heterogeneous catalysis. 　(4 marks)

3　Suggest why leaded petrol must not be used in a car fitted with a catalytic converter. 　(1 mark)

Measuring rate of reaction

'Rate of reaction' is a measure of how quickly a product is made, or a reactant is used up. To determine the rate of a reaction you need to measure the concentration of the reactant or product in question, or, a property that changes during the reaction and that is proportional to the concentration of the substance.

$$rate\ of\ reaction = \frac{Change\ in\ property}{time\ taken}$$

The unit of rate of reaction is $mol\,dm^{-3}\,s^{-1}$.

Measuring volumes of gases evolved

If a reaction produces a gas that can be collected, the volume produced can be used to follow the reaction rate. This can be done using a gas syringe, or by collecting the gas over water. Collecting gas over water is only suitable for gases that are insoluble in water, such as H_2 or O_2 (but not CO_2).

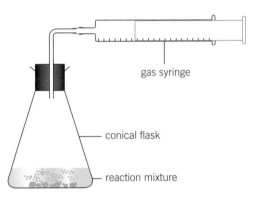

▲ **Figure 1** *Measuring rate of reaction using a gas syringe*

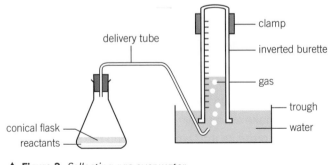

▲ **Figure 2** *Collecting gas over water*

Measuring mass changes

Another way of measuring the rate of a reaction that produces a gas is by recording the mass lost (in the form of gas) from the reaction.

pH measurement

Some reactions involve $H^+(aq)$ or $OH^-(aq)$ ions as a reactant, or produce them as a product. During these reactions, the concentration of the ions will change and the pH will consequently change. Measuring the rate of change of pH of the reaction mixture is a way of following the rate of reaction.

Colorimetry

A colorimeter measures the change in colour of a reaction. Many reactions result in a change of colour and the changing intensity of colour can be followed using a colorimeter.

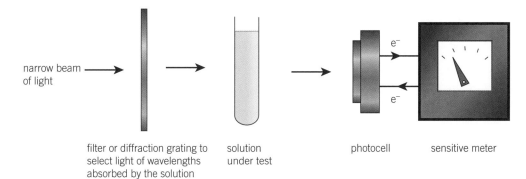

▲ **Figure 3** *A simple colorimeter*

Chemical analysis

Chemical analysis involves taking samples of the reaction mixture at regular intervals, and 'quenching' it before analysis. Quenching stops the reaction in the sample, or significantly slows it down.

For example, reactions involving an acid catalyst can be quenched by the addition of sodium hydrogen carbonate which neutralises the acid catalyst, effectively stopping the reaction.

Other quenching techniques include suddenly decreasing the temperature using an ice bath, or adding a large quantity of distilled water to dilute the substances. Both of these techniques decrease the rate of reaction significantly allowing time for the sample to be analysed by titration or other means.

1 What is the best explanation of why increasing the concentration increases the rate of reaction?

    A The particles are closer together, so they collide less frequently.

    B The particles are further apart, so there are fewer collisions.

    C The particles are closer together, so they collide more frequently.

    D The particles are further apart, so there are more collisions.   *(1 mark)*

2 Which of the following methods of measuring rate of reaction would be suitable for the reaction: $CuSO_4 + Zn \rightarrow ZnSO_4 + Cu$?

    A colorimetry   **B** measuring pH changes

    C measuring mass changes   **D** measuring volume changes   *(1 mark)*

3 What is the best definition of activation enthalpy?

    A The minimum kinetic energy required for a reaction to occur.

    B The energy released by a reaction.

    C The energy absorbed by the particles during a reaction.

    D The proportion of molecules with sufficient energy to react.   *(1 mark)*

4 How do catalysts increase the rate of reaction?

    A They decrease the activation energy of the reaction.

    B They increase the activation energy of the reaction.

    C They increase the energy of the colliding particles.

    D They provide an alternative pathway of lower activation energy. *(1 mark)*

5 The following stages explain the action of heterogeneous catalysis:

    1 bonds in the reactant break

    2 new bonds form

    3 reactant molecules are adsorbed onto the surface of the catalyst

    4 the products are released from the catalyst surface and diffuse away

    What is the correct order of these stages?

        A 1 – 3 – 4 – 2,   **B** 3 – 2 – 1 – 4

        C 4 – 2 – 1 – 3,   **D** 3 – 1 – 2 – 4   *(1 mark)*

6 In a petrochemical refinery, the reforming process converts straight-chain alkanes into ring compounds. Hot alkane vapours are passed over a catalyst. The catalyst used in the process is platinum metal, which is finely dispersed on a solid support of aluminium oxide.

    a Name the type of catalyst used in the reforming process.   *(1 mark)*

    b Explain why high temperatures are used.   *(2 marks)*

    c Explain why the catalyst is finely dispersed.   *(2 marks)*

    d Explain why the catalyst speeds up the rate of reaction.   *(2 marks)*

    e Describe how the catalyst works.   *(4 marks)*

7 Nitrogen monoxide, NO, is one of a number of radicals that catalyse the breakdown of ozone in the stratosphere.

    a Explain whether NO is a heterogeneous or homogeneous catalyst.

                                                   (1 mark)

    b Explain how NO molecules can act as catalysts for the breakdown of ozone, referring to intermediates, activation energy, and reaction pathways.   *(3 marks)*

    c Explain why the breakdown of ozone occurs faster in the hot, top layer of the stratosphere.   *(3 marks)*

11.1 Periodicity

Specification reference: EL(f), EL(m), EL(n), [EL(q)]

Periodicity

Periodicity is exhibited when there is a *regular pattern* in a property as you go across a period and the regular pattern is *repeated* in other periods.

Melting point – an example of periodicity

The pattern is for the melting point to increase and then decrease across a period. This pattern is repeated in more than one period. There is periodicity in melting points.

Melting point increases and then decreases across Period 2 (atomic numbers 3–10). There is the same increase and decrease in Period 3 (atomic numbers 11–18).

Metals are on the left-hand side of a period and have metallic bonding, with high melting points. Some Group 4 elements have covalent network structures and so tend to have high melting points. Other non-metals found on the right-hand side of a period have simple molecular structures and so tend to have lower melting points.

How does ionisation enthalpy vary across a period?

First ionisation enthalpy

The first ionisation enthalpy is the energy needed to remove one electron from each of *one mole* of *isolated gaseous* atoms of an element. *One mole* of *gaseous ions* with one positive charge is formed:

$$X(g) \rightarrow X^+(g) + e^-$$

Figure 1 shows the periodic pattern of first ionisation enthalpy for elements 1–56.

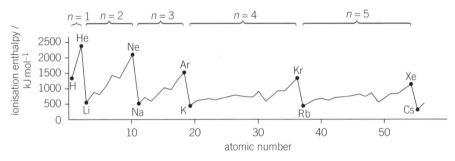

▲ **Figure 1** *First ionisation enthalpy of elements 1–56*

You should notice the following patterns:

- The general trend is one of *increasing* first ionisation enthalpy across each period.
- The elements at the peaks are all in Group 0. It is difficult to remove an electron from these atoms with full outer shells and the elements are all very unreactive.
- The elements at the troughs are all in Group 1. These atoms have only one outer shell electron and are relatively easy to ionise. They are all very reactive elements.

Ionisation enthalpies and reactivity

s-block elements are more reactive than p-block elements, because the formation of an M^+ or M^{2+} ion only requires input of energy equivalent to the first ionisation enthalpy (for M^+) or first and second ionisation enthalpy (for M^{2+}). Loss of electrons from p-block elements is more difficult.

Electronic configurations of s- and p-block elements

Group 1 and 2 elements all have one or two electrons, respectively, in their outermost sub-shell, which is an s-orbital. These are known as the s-block elements. For example, magnesium is $1s^2 2s^2 2p^6 3s^2$.

Group 3, 4, 5, 6, 7, and 0 elements all have three, four, five, six, seven, or eight electrons, respectively, in their outermost sub-shell, which are p-orbitals. They are known as the p-block elements. For example, selenium is $1s^2 2s^2 2p^6 3s^2 3p^6 4s^2 3d^{10}\,\mathbf{4p^4}$.

Elements in the d-block and f-block have electronic configurations where the outermost sub-shell is a d-orbital or an f-orbital. For example, nickel is in the d-block and is $1s^2 2s^2 2p^6 3s^2 3p^6 4s^2\,\mathbf{3d^8}$.

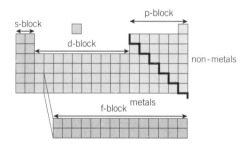

▲ **Figure 2** *s-, p-, d-, and f-blocks of the periodic table*

Summary questions

1 Predict the melting point of fluorine given the data below
   about other halogens.   (*1 mark*)

Element	Melting point (°C)
Fluorine	?
Chlorine	−102
bromine	7
Iodine	114

2 Name the elements which have these electronic
   configurations.   (*3 marks*)
   a $1s^2\, 2s^2\, 2p^5$
   b $1s^2\, 2s^2\, 2p^6\, 3s^2\, 3p^6\, 4s^2$
   c $1s^2\, 2s^2\, 2p^6\, 3s^2\, 3p^6\, 4s^2\, 3d^{10}$

3 Classify the elements in Question 2 as s-, p-, or
   d-block elements.   (*3 marks*)

4 Given below are the values of the first four ionisation enthalpies of
   calcium and aluminium.

   1st I.E. $(Mg) = 736\,kJ\,mol^{-1}$ 1st I.E. $(Al) = 577\,kJ\,mol^{-1}$
   2nd I.E. $(Mg) = 1450\,kJ\,mol^{-1}$ 2nd I.E. $(Al) = 1820\,kJ\,mol^{-1}$
   3rd I.E. $(Mg) = 7740\,kJ\,mol^{-1}$ 3rd I.E. $(Al) = 2740\,kJ\,mol^{-1}$
   4th I.E. $(Mg) = 10\,500\,kJ\,mol^{-1}$ 4th I.E. $(Al) = 11\,600\,kJ\,mol^{-1}$

   a Use this information to explain why magnesium forms
      Mg^{2+} ions, whilst aluminium forms Al^{3+} ions.   (*2 marks*)
   b Calculate the total energy required to form Mg^{2+} and Al^{3+},
      and explain why magnesium is a more reactive element
      than aluminium.   (*3 marks*)

Revision tip

In these equations, M is any Group 2 metal.

Revision tip

In these equations, the state of $M(OH)_2$ could be (s) or (aq), depending on the solubility of the metal hydroxide.

Revision tip

Remember nitric acid is monobasic, so two moles of HNO_3 react with each mole of the oxide.

Revision tips

Group 2 elements have similar chemical reactions, as they all have two electrons in their outer shell.

Group 2 metals are less reactive than Group 1 metals. For example, Na reacts with cold water, whereas Mg reacts with steam.

Revision tips

Oxides and hydroxides of Group 2 metals are basic.

The reaction of oxides and hydroxides with acids are very similar.

Reactions of the elements in Group 2

The Group 2 elements are Be, Mg, Ca, Sr, Ba, and Ra.

The metals react with water to give the metal hydroxide and hydrogen:

$$\text{metal} + \text{water} \rightarrow \text{metal hydroxide} + \text{hydrogen}$$
$$\mathbf{M}(s) + 2H_2O(l) \rightarrow \mathbf{M}(OH)_2(s) + H_2(g)$$

As is typical for metals, the reactions become more vigorous as you go down the group.

The oxides

The oxides react with water to produce an alkaline solution of the hydroxide:

$$\text{metal oxide} + \text{water} \rightarrow \text{metal hydroxide}$$
$$\mathbf{M}(s) + H_2O(l) \rightarrow \mathbf{M}(OH)_2(s)$$

The oxides react with acids and therefore act as bases, accepting a proton:

$$\text{metal oxide} + \text{acid} \rightarrow \text{salt} + \text{water}$$
$$\mathbf{M}O(s) + H_2SO_4(aq) \rightarrow \mathbf{M}SO_4(aq) + H_2O(l)$$
$$\mathbf{M}O(s) + 2HNO_3(aq) \rightarrow \mathbf{M}(NO_3)_2(aq) + H_2O(l)$$

The hydroxides

The hydroxides become more soluble as you go down the group. The solutions produced are alkaline, since the solutions contain $OH^-(aq)$ (pH > 7).

The hydroxides react with acids to produce a salt and water:

$$\text{metal hydroxide} + \text{acid} \rightarrow \text{salt} + \text{water}$$
$$\mathbf{M}(OH)_2(s) + 2HCl(aq) \rightarrow \mathbf{M}Cl_2(aq) + 2H_2O(l)$$

The carbonates

The carbonates become less soluble as you go down the group.

The carbonates undergo thermal decomposition on heating to give the metal oxide and carbon dioxide:

$$\text{metal carbonate} \rightarrow \text{metal oxide} + \text{carbon dioxide}$$
$$\mathbf{M}CO_3(s) \rightarrow \mathbf{M}O(s) + CO_2(g)$$

Thermal stability increases down the group. This means that barium carbonate breaks down at a higher temperature than magnesium carbonate.

Common misconception: Thermal decomposition

Avoid incorrectly stating that barium carbonate will be more likely to break down on heating as barium is more reactive.

The M^{2+} ions get larger as you go down Group 2, so their charge density is lower. Ions such as Ba^{2+}, with a low charge density, polarise the carbonate ion less than ions such as Mg^{2+}, which has a high charge density. The more polarised the carbonate ion, the more likely it is to break up during thermal decomposition into an oxide ion and CO_2.

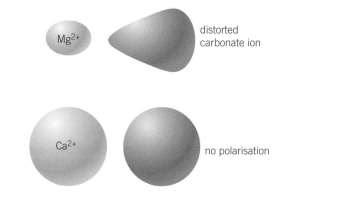

distorted carbonate ion

no polarisation

Summary questions

1 Write a balanced equation with state symbols for the reaction of strontium with water.   *(1 mark)*

2 Write a balanced equation, with state symbols, for the thermal decomposition of calcium carbonate.   *(1 mark)*

3 Which one in each pair is the most soluble?
  a $Mg(OH)_2$ or $Ba(OH)_2$ *(1 mark)*
  b $MgCO_3$ or $BaCO_3$.   *(1 mark)*

4 Write a balanced equation, with state symbols, for the reaction of:
  a magnesium oxide with hydrochloric acid.   *(1 mark)*
  b magnesium hydroxide and sulfuric acid.   *(1 mark)*

Properties of the halogens
Physical properties

You need to be able to recall the following physical properties of the halogens:

	Fluorine (F_2)	Chlorine (Cl_2)	Bromine (Br_2)	Iodine (I_2)
Appearance and state at room temperature	pale yellow gas	green gas	dark red liquid	shiny grey/black solid
Volatility	gas	gas	liquid quickly forms brown gas on warming	sublimes on warming to give a purple vapour
Solubility in water	reacts with water	slightly soluble to give pale green solution	slightly soluble to give orange-yellow solution	barely soluble, gives a brown solution
Solubility in organic solvents	soluble	soluble to give a pale green solution	soluble to give a orange/brown/red solution	soluble to give a violet solution

Common misconception: Formulae of the halogens

Remember that the elements are diatomic molecules, so should be written as F_2, etc.

All the halogens exist as elements as *diatomic* non-polar molecules (F_2, Cl_2, Br_2 and I_2). The *intramolecular* bonds are covalent and the *intermolecular* bonds are instantaneous dipole–induced dipole bonds. Fluorine is the most volatile halogen, as it has the smallest molecules with the fewest electrons. As the size of the molecule and number of electrons increases, so does the strength of the intermolecular bonds. This explains why the physical state of the halogens changes from gas to liquid to solid as you go down the group.

The elements become darker in colour as you go down the group.

Reactivity and redox reactions of the halogens

Halogens are all reactive. They tend to remove electrons from other elements – they are oxidising agents:

$$X_2 + 2e^- \rightarrow 2X^-$$

Halogen oxidation state 0 is reduced to oxidation state −1.

Halogens and halide ions undergo displacement reactions. For example, when a solution containing chlorine is added to a solution containing iodide ions, a brown colour appears as iodine is produced:

$$Cl_2(aq) + 2I^-(aq) \rightarrow I_2(aq) + 2Cl^-(aq)$$

This occurs because chlorine is a stronger oxidising agent than iodine. Chlorine causes the iodine to be oxidised (the oxidation state of iodine changes from −1 to 0 in the half-equation $2I^- \rightarrow I_2 + 2e^-$). Chlorine has been reduced (its oxidation state changes from 0 to −1 in the half-equation $Cl_2 + 2e^- \rightarrow 2Cl^-$).

- Fluorine is the strongest oxidising agent in Group 7. Fluorine displaces chlorine, bromine, and iodine.
- Chlorine displaces bromine and iodine.

- Bromine displaces iodine.
- Iodine does not displace bromine, chlorine, or fluorine.

An organic solvent such as hexane can be added to detect the halogen present at the end of the reaction. For example, in the reaction of chlorine with iodide ions shown above, the hexane layer would be violet due to the presence of iodine.

Precipitation reactions with silver ions

The general reaction is:

$$Ag^+(aq) + X^-(aq) \rightarrow AgX(s)$$

Chloride ions give a white precipitate of silver chloride, AgCl, which is soluble in dilute ammonia. Bromide ions give a cream precipitate of silver bromide, AgBr, which is soluble in concentrated ammonia. Iodide ions give a yellow precipitate of silver iodide, AgI, which is insoluble in ammonia.

These precipitation reactions can be used as tests to identify halides in solution.

Revision tip

Look back at the colours of halogens in organic solvents in the table above.

Revision tip

Remember to include state symbols in precipitation reactions.

Key term

Precipitate: An insoluble solid formed when two ionic solutions react together.

Summary questions

1   Describe how the appearance of iodine differs depending
    on the conditions present.   (4 marks)

2   Write an ionic equation for the reaction of potassium iodide
    solution with silver nitrate. State what you would see.   (2 marks)

3   Chlorine gas is bubbled through potassium bromide solution.
    a   Write an ionic equation for the reaction that occurs.   (1 mark)
    b   Describe and explain what you would see.   (2 marks)
    c   Describe and explain what you would see if cyclohexane
        (an organic solvent) was added after the reaction.   (2 marks)
    d   Explain why no reaction would occur if bromine were
        added to potassium chloride solution.   (1 mark)

1 How are elements arranged in the modern periodic table?

   A increasing relative atomic mass

   B increasing atomic radius

   C increasing reactivity

   D increasing atomic number   (1 mark)

2 Which element has the electronic configuration $1s^2\ 2s^2\ 2p^6\ 3s^2\ 3p^6\ 4s^2\ 3d^2$?

   A calcium   B scandium

   C gallium   D titanium   (1 mark)

3 Which of the following equations represents the third ionisation enthalpy of silicon?

   A $Si(g) \rightarrow Si^{3+}(g) + e^-$ B $Si(s) \rightarrow Si^{3+}(s) + 3e^-$

   C $Si^{2+}(s) \rightarrow Si^{3+}(s) + 3e^-$ D $Si^{2+}(g) \rightarrow Si^{3+}(g) + e^-$ (1 mark)

4 Which elements appear at the peaks of a plot of first ionisation energy for successive elements in the periodic table?

   A Group 0 elements   B Group 1 elements

   C Group 7 elements   D Transition elements   (1 mark)

5 Which statement(s) are true for Group 2 elements?

   1 The hydroxides become more soluble down the group.

   2 Group 2 elements are more reactive than Group 1 elements.

   3 The carbonates become more soluble down the group.

   A 1 only   B 1 and 2 only

   C 2 and 3 only   D 1, 2, and 3   (1 mark)

6 Which statement is true about thermal decomposition of Group 2 carbonates?

   A $MgCO_3$ is the most thermally stable Group 2 carbonate.

   B $BaCO_3$ will decompose more easily than $SrCO_3$ on heating.

   C The carbonate ion in $SrCO_3$ is more polarised than in $CaCO_3$.

   D Ca^{2+} has a higher charge density than Sr^{2+}.   (1 mark)

7 Which halogens will displace bromine?

   A fluorine, chlorine, and iodine

   B iodine only

   C fluorine and chlorine only

   D potassium bromide.   (1 mark)

8 In the following reaction: $Br_2(aq) + 2I^-(aq) \rightarrow I_2(aq) + 2Br^-(aq)$

   A bromine is oxidised

   B bromine is a weaker oxidising agent than iodine

   C iodine is an oxidising agent

   D iodine is oxidised.   (1 mark)

9 Which statement is correct about the volatility of halogens?

   A Fluorine is the most volatile as it has the weakest intermolecular forces.

   B Fluorine is the most volatile as it has the strongest intermolecular forces.

   C Fluorine is the least volatile as it has the weakest intermolecular forces.

   D Fluorine is the least volatile as it has the strongest intermolecular forces.   (1 mark)

10 Which substance is a yellow solid, insoluble in concentrated ammonia?

  A silver bromide

  B silver chloride

  C silver iodide

  D silver nitrate   (*1 mark*)

11 a Special precautions must be taken when transporting bromine.
      Emergency breathing apparatus and protective suits must be carried
      by the lorry driver. Describe two properties of bromine that
      make this necessary.   (*2 marks*)

  b Explain, in terms of intermolecular bonds, why bromine is a
      liquid at room temperature and pressure but chlorine is a gas. (*4 marks*)

12 Magnesium oxide has a very high melting point and is used
   for making furnace bricks. Explain this in terms of the charge
   densities of the ions involved.   (*2 marks*)

13 The Romans discovered that thermal decomposition of limestone,
   $CaCO_3$ can be used to make a product that neutralises acid soil.

  a Write an equation for the thermal decomposition of limestone. (*1 mark*)

  b Identify a Group 2 carbonate that decomposes more readily
      than $CaCO_3$. Explain why it decomposes more readily.   (*2 marks*)

  c Name the product of the thermal decomposition of limestone. (*1 mark*)

  d Show how the product in (c) reacts with an acid. Use the
      general symbol HA to represent any acid.   (*2 marks*)

Straight-chain alkanes

- have the general formula C_nH_{2n+2}
- name ends in -*ane*
- are **saturated** – all the bonds between carbon atoms are *single bonds*
- are **aliphatic** – they do not contain benzene rings.

Naming alkanes

- Choose the longest carbon chain and name it.
- Use prefixes in alphabetical order for any alkyl side chains.
- Use *di*, *tri*, *tetra* before the prefix if the side chains are identical.
- Show the position of any side chains by using numbers which are as low as possible.

Examples of alkane names

Put a comma between numbers and a hyphen between a number and a letter:

$CH_3CH(CH_3)CH_3$ is methylpropane

$CH_3CH(CH_3)CH_2CH_2CH_3$ is 2-methylpentane

$CH_3C(CH_3)_2CH_3$ is 2,2-dimethylpropane

$CH_3CH(CH_3)CH_2CH(CH_3)CH_2CH(CH_3)CH_3$ is 2,4,6-trimethylheptane.

Different types of formulae

Full structural formulae show all the bonds and all the atoms in the molecule. Shortened structural formulae abbreviate groups such as CH_3 or CH_2, but still clearly show the arrangement of groups of atoms in the molecule.

▼ **Table 1** *Different types of formulae*

Name	Molecular formula	Full structural formula	Shortened structural formula	Further shortened to
methane	CH_4	H—C—H (with H above and below)	CH_4	
ethane	C_2H_6	H—C—C—H (with H above and below each C)	$CH_3–CH_3$	CH_3CH_3
propane	C_3H_8	H—C—C—C—H (with H above and below each C)	$CH_3–CH_2–CH_3$	$CH_3CH_2CH_3$

Skeletal formulae use lines to represent carbon–carbon bonds. Hydrogen atoms attached to C atoms are not drawn in skeletal formulae. The skeletal formulae of butane and methylbutane are shown in Figures 1 and 2.

Key terms

Alkane: An alkane is a saturated hydrocarbon.

Structural formula: A representation of the atoms, bonds, and groups in a molecule.

Skeletal formula: A representation of a molecule using lines to represent C–C bonds.

▲ **Figure 2** *Skeletal formula of methylbutane*

▲ **Figure 1** *Skeletal formula of butane*

Structural and skeletal formulae do not accurately represent the 3-dimensional shapes of molecules. To get round this problem, chemists use wedge and dotted bonds:

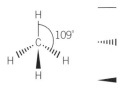

— represents a bond in the plane of the paper

⦙⦙⦙⦙⦙ represents a bond in a direction behind the plane of the paper

◀ represents a bond in a direction in front of the plane of the paper

▲ **Figure 4** *Three-dimensional structure of ethane*

▲ **Figure 3** *Three-dimensional structure of methane*

The carbon chain found in alkanes can be substituted by different functional groups such as −OH, −Cl, −COOH. This gives different homologous series – a group of molecules with the same functional group but different carbon chain lengths. Members of homologous series have chain lengths differing by a CH_2 unit.

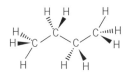

▲ **Figure 5** *Three-dimensional structure of butane*

Cycloalkanes

- have the general formula C_nH_{2n}
- name ends in *-ane*
- are **saturated** – all the bonds between carbon atoms are *single bonds*
- are *not* **aromatic** – they don't have a benzene ring
- are aliphatic.

Summary questions

1 Draw the skeletal formulae for 3-methylhexane and 2,2,4-trimethylheptane. *(2 marks)*

2 Give the molecular formula of (i) an alkane and (ii) a cycloalkane with 10 carbon atoms. *(2 marks)*

3 Describe what the molecular formulae of hexene, 2-methylpentene, and cyclohexane have in common. *(1 mark)*

Unsaturated hydrocarbons

Alkenes are a homologous series of hydrocarbons and have the following characteristics:

- the general formula C_nH_{2n}
- name ends in *-ene*
- are **unsaturated** – there are one (or more) *double bonds* between carbon atoms in a molecule
- are **aliphatic** – they don't have delocalised benzene ring structures.

Naming alkenes

The names of alkenes end in *-ene*. The number preceding the -ene indicates the position of the double bond. But-1-ene is $CH_3-CH_2-CH=CH_2$ and but-2-ene is $CH_3-CH=CH-CH_3$.

Shape of alkenes

All bond angles around the C=C double bond are 120°, since there are three groups of electrons around each carbon atom (two single bonds and one double bond). These groups of electrons repel each other as far as possible.

A C=C bond contains a sigma (σ) bond and a pi (π) bond. A σ-bond is an area of increased electron density between the carbon atoms. A π-bond consists of two areas of negative charge. One of these is above the line of the atoms, and the other is below.

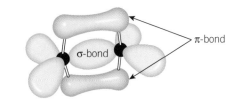

▲ **Figure 1** *Sigma- and pi-orbitals*

Reactions of alkenes

Alkenes undergo **electrophilic addition** reactions. In the following examples we will use ethene as a typical alkene. There are four electrons in the double bond of ethene – these give the region between the two carbon atoms a high negative charge density. Electrophiles are attracted to this negatively charged region in an alkene and accept a pair of electrons from the double bond at the start of the reaction. When electrophiles react, they accept a pair of electrons. A general scheme for the reactions is shown in Figure 2, using bromine as an electrophile.

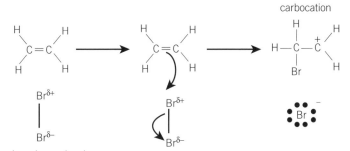

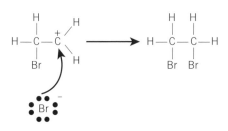

▲ **Figure 2** *Electrophilic addition*

- When a bromine molecule approaches an alkene, it becomes **polarised.**
- The electrons in the bromine molecule are repelled back along the molecule.
- The electron density is *unequally distributed.*
- The bromine atom nearest to the alkene becomes slightly positively charged.
- It now acts as an **electrophile.** A pair of electrons from the alkene moves towards the slightly charged bromine atom and a C–Br bond is formed.
- The carbon species is now positively charged; it is a **carbocation.**
- The other bromine, now negatively charged, moves in rapidly to make another bond.

The overall process is addition by an electrophile across a double bond – it is **electrophilic addition.** This model for the mechanism for electrophilic addition (*via* a carbocation) can be supported by experimental evidence. If chloride ions, Cl-, are present when ethene reacts with bromine, the molecule $BrCH_2CH_2Cl$ forms, as well as the expected $BrCH_2CH_2Br$. This is because both chloride and bromide ions can attack the intermediate carbocation.

Alkenes can react with a number of electrophiles.

Electrophile	Product	Conditions
Br_2	CH_2BrCH_2Br 1,2-dibromoethane	room temperature and pressure
HBr(aq)	CH_3CH_2Br bromoethane	aqueous solution, room temperature and pressure
H_2O (H–OH)	CH_3CH_2OH ethanol	phosphoric acid / silica at 300 °C / 60 atm or with conc. H_2SO_4, then H_2O at 1 atm
H_2	CH_3CH_3 ethane	Pt catalyst, room temperature and pressure, or Ni catalyst at 150 °C / 5 atm

Addition polymerisation

Alkenes are the basic hydrocarbon units of many polymers:

- –A–A–A–A–A–A– polymers are made from one type of alkene monomer
- –A–B–A–B–A–B– polymers can be made from more than one type of alkene monomer.

Key terms

Alkene: Alkenes are unsaturated hydrocarbons containing one or more C=C bonds.

Electrophiles: Positive ions or molecules with a partial positive charge on one of the atoms.

Electrophilic addition: A reaction in which an electrophile joins onto an alkene. No atoms are removed from the alkene.

Polarised: A covalent bond is said to be polarised if the electrons are unevenly distributed between the atoms.

Common misconception: Carbocations

The correct pronunciation of carbocation is 'carbo – cat – ion'.

Revision tip

When bromine approaches an alkene it becomes polarised, because the electrons in the Br–Br bond are repelled from the electron-dense C=C bond.

Key term

Carbocation: A carbon-containing positive ion, formed as an intermediate in electrophilic addition.

Revision tip

Make sure you know the reaction conditions for all the reactions in the table.

Revision tip

Bromine is used as a test for unsaturation. The brown bromine becomes decolourised.

Under the right conditions, alkenes can undergo addition polymerisation. The small *unsaturated* starting molecules are called **monomers** and they join together to form a long chain *saturated* **polymer**. No other product is formed.

$$CH_2=CHCH_3 + CH_2=CHCH_3 + CH_2=CHCH_3 \rightarrow$$
$$-CH_2-CH(CH_3)-CH_2-CH(CH_3)-CH_2-CH(CH_3)-$$

propene (monomers)   *poly(propene)* (polymer)

This may be written as:

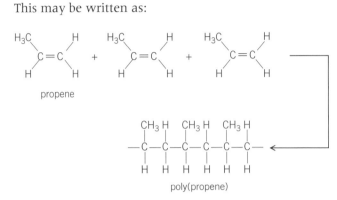

▲ **Figure 3** *Addition polymerisation*

The repeat unit is written as:

▲ **Figure 4** *Repeat unit of poly(propene)*

The polymer is named by putting the name of the monomer in brackets and prefixing with 'poly', for example, choroethene monomer gives poly(chloroethene) as the polymer. Note, however, that the polymer is *not* an alkene.

Summary questions

1 Draw the full structural, shortened structural, and skeletal formulae of pent-1-ene. *(3 marks)*

2 Draw a skeletal formula of the product when $H_2(g)$ reacts with pent-2-ene. *(1 mark)*

3 Draw the mechanism of the electrophilic addition of HBr to but-2-ene. *(4 marks)*

4 Draw the structure of the polymer formed when the monomer CHCl=CHCl undergoes addition polymerisation, showing three repeat units. *(2 marks)*

12.3 Structural isomerism and E/Z isomerism

Specification reference: DF(s), DF(t)

Structural isomers

Chain isomerism, in which the chain lengths are different because of branching, is often seen in alkanes. Butane has a four-carbon chain, whereas methylpropane has three carbon atoms in its longest chain, but both have the molecular formula C_4H_{10}.

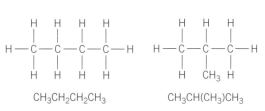

▲ **Figure 1** Isomers of C_4H_{10}

Position isomerism, in which the same functional group appears in different positions, is often seen in alcohols. For example, propan-1-ol, $CH_3CH_2CH_2OH$, and propan-2-ol, $CH_3CH(OH)CH_3$, both have the molecular formula C_3H_8O.

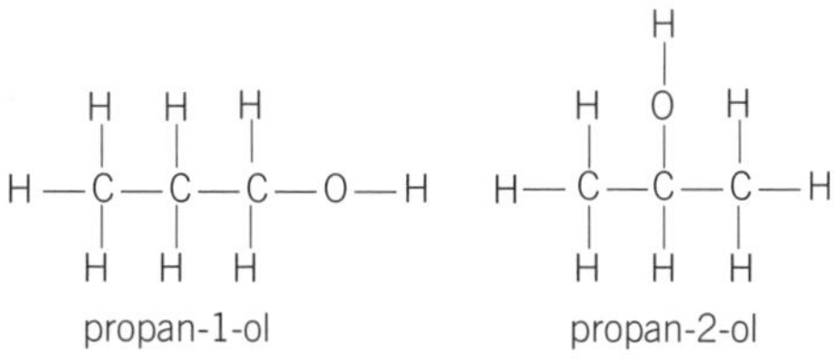

▲ **Figure 2** Isomers of C_3H_8O

Functional group isomerism, in which the molecular formulae are the same but the functional groups are different, can be seen in alcohols and ethers, for example ethanol, C_2H_5OH, and methyoxymethane, CH_3OCH_3.

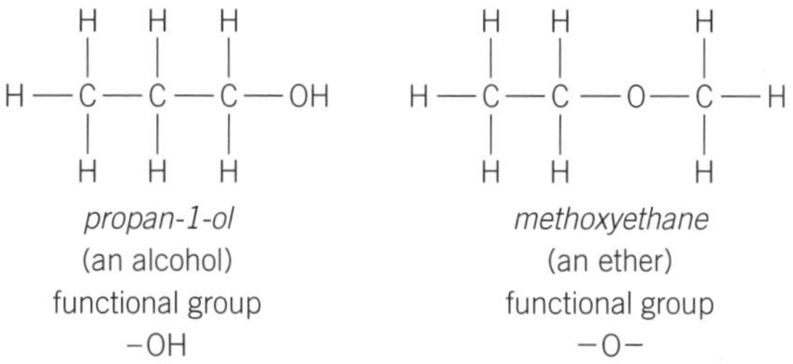

propan-1-ol (an alcohol) functional group −OH

methoxyethane (an ether) functional group −O−

▲ **Figure 3** Functional group isomerism

Stereoisomerism

E/Z isomerism is one type of stereoisomerism. In stereoisomerism, the atoms are bonded in the same order, but are arranged differently in space in each isomer.

The isomer of but-2-ene that has the two methyl groups on the same side of the double bond is called the *Z* isomer, that is *Z*-but-2-ene. It is also called *cis*-but-2-ene. The isomer that has the two methyl groups on opposite sides of the double bond is called the *E* isomer, that is *E*-but-2-ene. It is also called *trans*-but-2-ene.

The reason why these two isomers exist is that to turn one form into the other you need to break one of the bonds in the carbon-carbon double bond. There is not enough energy at room temperature to enable this to occur. As a result, interconversion of the two isomers does not occur.

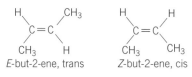

E-but-2-ene, trans   *Z*-but-2-ene, cis

▲ **Figure 4** *E/Z* isomers of but-2-ene

Summary questions

1 Draw the isomers of $C_2H_4Br_2$. (*2 marks*)

2 Draw the skeletal formulae of 2-methylpropan-2-ol and 2-methylpropan-1-ol and explain whether or not they are isomers. (*3 marks*)

3 Explain why 2-methylpropene does not exhibit E/Z isomerism. (*1 mark*)

1 What is the name of the molecule $CH_3CH(CH_3)CH(CH_3)CH_3$?

   A hexane

   B 2,3-dimethylhexane

   C 2,3-dimethylbutane

   D butane-2,3-dimethyl   (*1 mark*)

2 Skeletal formulae

   A use straight lines to represent C−C bonds

   B show hydrogen atoms

   C cannot represent functional groups

   D accurately represent the 3D shapes of molecules   (*1 mark*)

3 Ethanol and methoxymethane

   A are examples of position isomerism

   B both have the molecular formula C_2H_5OH

   C are examples of functional group isomerism

   D are both alcohols   (*1 mark*)

4 Which molecule does **not** have the molecular formula C_5H_{10}?

   A pent-1-ene

   B pentane

   C cyclopentane

   D 2-methylbut-2-ene   (*1 mark*)

5 Which statements are true about alkanes and alkenes?

       1 Alkanes and alkenes are hydrocarbons.

       2 Alkanes and alkenes can both undergo addition polymerisation.

       3 Alkenes with one double C=C bond have the same molecular formula as the cycloalkane with the same number of carbons.

   A 1 only

   B 2 and 3 only

   C 1 and 3 only

   D 1, 2, and 3   (*1 mark*)

6 A hydrocarbon has an M_r of 58. It reacts with bromine water and can produce an addition polymer. It has E/Z isomers. Identify the hydrocarbon, explaining the significance of each of its properties.   (*4 marks*)

7 Decane is a hydrocarbon which is found in petrol.

   a Give the molecular formula of decane.   (*1 mark*)

   b Explain why all the bond angles in decane are 109°.   (*3 marks*)

   Decane can undergo catalytic cracking to produce two shorter molecules, an alkene, C_3H_6, and an alkane, C_7H_{14}.

   c Name C_3H_6 and C_7H_{14}.   (*2 marks*)

   d Name the *types* of isomerism that C_3H_6 and C_7H_{14} can show. (*2 marks*)

   C_3H_6 was reacted with aqueous bromine.

   e Name the type of reaction that occurs.   (*1 mark*)

   f Give the name of the main product.   (*1 mark*)

   g Explain why the main product is contaminated with $CH_2BrCH(OH)CH_3$. (*1 mark*)

13.1 Naming organic compounds: Alkanes, haloalkanes, and alcohols

Specification reference: DF(m), OZ(j), WM(b)

Alkanes

Alkanes are named by looking for the longest chain of carbon atoms. The name ends in -ane and depends on the number of carbon atoms in the chain: *meth*ane, *eth*ane, *prop*ane, *but*ane, *pent*ane, etc. Branched molecules are named based on the parent alkane. For example, CH_3 is a *methyl* group and C_2H_5 is an *ethyl* group. Numbers are used to show the position of the group, using the lowest numbers possible. For example, $CH_3CH_2CH_2CH(CH_3)CH_3$ is 2-methylpentane.

Haloalkanes

The homologous series of the halogenoalkanes is an alkane series with hydrogen atoms substituted by one or more halogen atoms. They are often shown as R—Hal, where Hal could be F, Cl, Br, or I.

Naming haloalkanes

The alkane chain name is *prefixed* with the name of the halogen. The halogens are listed in alphabetical order, with a number indicating the position of each. For example, C_2H_5Cl is chloroethane, $CH_3CHBrCH_3$ is 2-bromopropane, and CH_2BrCH_2I is 1-bromo-2-iodoethane.

Alcohols

Alcohols all contain the –OH functional group and their names end in -ol. They may be classfied as primary, secondary, or tertiary. It is the position of the –OH group which determines the classification.

Type of alcohol	Position of –OH group	General formula	Example
primary	–OH bonded to a carbon bonded to *one* other carbon atom	RCH_2OH	butan-1-ol
secondary	–OH bonded to a carbon bonded to *two* other carbon atoms	$RCH(OH)R$	butan-2-ol
tertiary	–OH bonded to a carbon bonded to *three* other carbon atoms	$R_2C(OH)R$	2-methylpropan-2-ol

Summary questions

1 Describe the differences between:
   a 2-methylpentane and 3-methylpentane; *(1 mark)*
   b 2,2-dimethylbutane and 2,3-dimethylbutane; *(1 mark)*
   c 3-methylhexane and 3-methylheptane. *(1 mark)*

2 Explain why these alkanes are incorrectly named:
   a 1-methylpropane *(1 mark)*
   b 2-ethylbutane *(1 mark)*
   c 2,3-methylbutane. *(1 mark)*

3 Draw and name primary, secondary, and tertiary alcohols each containing four carbon atoms. *(3 marks)*

4 For each pair of haloalkanes, choose the correct name:
   a 1-bromo-2-chloroethane or 1-chloro-2-bromoethane *(1 mark)*
   b 2-chloro-2-iodopropane or 2-iodo-2-chloropropane *(1 mark)*
   c 1,2,3-iodopropane, triiodopropane, or 1,2,3-triiodopropane. *(1 mark)*

Physical properties of haloalkanes

The boiling points increase with a heavier halogen atom (R–I > R–F) or with increasing numbers of halogen atoms (CCl_4 > CH_2Cl_2). As the halogen introduced is larger or the number of halogen atoms increases, the overall number of electrons increases. This increases the instantaneous dipole–induced dipole bonds. With stronger intermolecular bonds, more energy is needed to pull the molecules apart from each other, so the boiling point is higher.

Bond enthalpies and reactivity of haloalkanes

Bond within the molecule	Bond strength	Reactivity
C–F	decreasing strength	increasing reactivity
C–Cl		
C–Br	↓	↓
C–I		

The C–Hal bond becomes weaker as the size of the halogen atom increases. This makes the bond easier to break and the compounds become more reactive. Although the C–F bond is the most polar, fluoroalkanes are very unreactive. This shows that it is bond strength rather than bond polarity that has the greatest effect on the reactivity of haloalkanes.

- Fluoro- compounds are very unreactive.
- Chloro- compounds are reasonably stable in the troposphere and can react to produce chlorine radicals that deplete ozone.
- Bromo- and iodo-compounds are reactive and so are useful as intermediates in chemical synthesis.

Reactions of haloalkanes

Homolytic fission (forming radicals):

$$R–Hal \rightarrow R\bullet + Hal\bullet$$

The C–I bond most easily undergoes homolytic fission, as it has the lowest bond enthalpy.

Heterolytic fission

$$R–Hal \rightarrow R^+ + Hal^-$$

The carbon–halogen bond breaks to give ions. If the polar C–Hal bond is broken completely, a negative halide ion moves away, leaving the C group positively charged. This is now a **carbocation**.

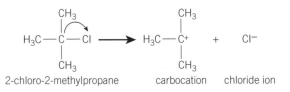

2-chloro-2-methylpropane   carbocation   chloride ion

▲ **Figure 1** *Heterolytic fission in 2-chloro-2-methylpropane*

Substitution reactions:

$$R–Hal + X^- \rightarrow R–X + Hal^-$$

Synoptic link

You learned about instantaneous dipole–induced dipole bonds in Topic 5.3, Bonds between molecules.

Revision tip

Bond polarity suggests that the C–F bond should be the easiest to break as it is the most polar, and that the C–I bond should be the hardest to break as it is least polar.

Revision tip

Bond strength indicates that the C–F bond should be the hardest to break as it is the strongest, and that the C–I bond should be the easiest to break as it is weakest.

Revision tip

As iodoalkanes are the most reactive the C–I bond is the easiest to break and therefore bond strength, not bond polarity, is the determining factor.

Key term

Homolytic fission: A type of bond breaking where each atom in a covalent bond gains one electron, forming radicals.

Key term

Radical: A species with an unpaired electron.

Common misconception: Radicals

When describing radicals, avoid saying they contain a lone electron.

In this case, the C−Hal bond breaks and the halogen atom is replaced by another functional group. Since the halogen is replaced by a nucleophile these reactions are called **nucleophilic substitution** reactions.

The mechanism of nucleophilic substitution is shown in Figure 2.

▲ **Figure 2** *General reaction mechanism for nucleophilic substitution*

Nucleophile	Equation	Product	Reaction conditions
H_2O	$R-Hal + H_2O \rightarrow$ $R-OH + H^+ + Hal^-$	alcohol	heat under reflux: this is sometimes called hydrolysis
OH^-	$R-Hal + OH^- \rightarrow$ $R-OH + Hal^-$	alcohol	heated under reflux with $NaOH(aq)$, with ethanol as solvent
$:NH_3$	$R-Hal + NH_3 \rightarrow$ $R-NH_2 + Hal^- + H^+$	amine	the haloalkane is heated with concentrated ammonia solution in a sealed tube

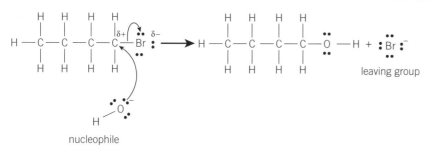

▲ **Figure 3** *Nucleophilic substitution reaction between hydroxide ions and 1-bromobutane*

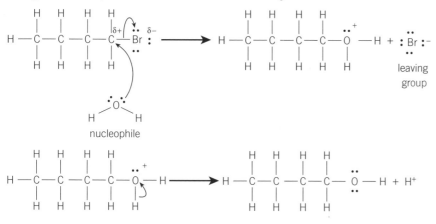

▲ **Figure 4** *Nucleophilic substitution reaction between water molecules and 1-bromobutane – this occurs in two steps. The ion resulting from step 1 loses H⁺ to form an alcohol. This is known as a hydrolysis reaction.*

Revision tip

Conditions for homolytic fission are gas phase with high temperatures or the presence of UV radiation (e.g. in the stratosphere).

Key term

Heterolytic fission: A type of bond breaking where both electrons in a covalent bond go to one atom, forming ions.

Revision tip

Conditions for heterolytic fission are: dissolved in a polar solvent such as an ethanol / water mixture.

Key term

Carbocation: an ion with a positively-charged carbon atom.

Revision tip

A full headed arrow indicates the movement of a pair of electrons. A half-headed arrow indicates the movement of one electron.

Key term

Nucleophile: A molecule, or negatively charged ion, with a lone pair of electrons that it can donate to a positively charged atom to form a covalent bond.

Summary questions

1 Draw the skeletal formula for $CH_3CH_2CHBrCCl_2CHICH_3$ and name the compound. *(1 mark)*
2 Which in each of the following pairs of molecules has the higher boiling point? Explain your reasoning.
   a CH_3Br and CH_3F
   b CH_3Br and CBr_4. *(4 marks)*

3 Explain why iodoethane is more reactive than fluoroethane. *(1 mark)*
4 Explain the term *nucleophile* and give two examples. *(3 marks)*

5 Give the conditions needed for ammonia to react with a haloalkane. *(1 mark)*
6 Draw the mechanism for the nucleophilic substitution reaction between the hydroxide ion and 1-iodopropane. *(4 marks)*

Primary, secondary, and tertiary alcohols

Primary alcohols have an –OH group bonded to a carbon bonded to one other carbon atom. Secondary alcohols have an –OH group bonded to a carbon bonded to two other carbon atoms. Tertiary alcohols have an –OH group bonded to a carbon bonded to three other carbon atoms.

Oxidation of alcohols

The –OH group can be oxidised using acidified potassium dichromate(VI), $K_2Cr_2O_7$. The –OH group is oxidised to a carbonyl group. At the same time, the $Cr_2O_7^{2-}$ ion, which is orange, is reduced to Cr^{3+}(aq), which is green. During the oxidation reaction, two hydrogen atoms are removed – one from the oxygen atom and one from the carbon atom. The reaction conditions for the oxidation of alcohols are heating the alcohol under reflux with excess acidified potassium (or sodium) dichromate(VI) solution.

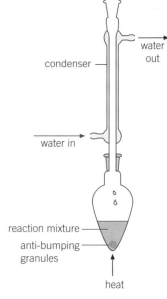

▲ **Figure 1** *Heating under reflux*

The products of oxidation

The product depends on the type of alcohol used. With a primary alcohol, an aldehyde is produced, which can oxidise further to give a carboxylic acid. The colour of the reaction mixture changes from orange to green as the dichromate(VI) ion is reduced. If the aldehyde is required it can be distilled out of the reaction mixture as it is produced, in order to prevent further oxidation. If the carboxylic acid is required, the mixture is heated under reflux with excess potassium dichromate(VI) solution. With a secondary alcohol, a ketone is produced and no further oxidation occurs. The colour of the reaction mixture changes from orange to green. Tertiary alcohols do not undergo oxidation with acidified potassium dichromate(VI), because they do not have a hydrogen atom on the carbon to which –OH is attached. The colour of the reaction mixture does not change, but remains orange.

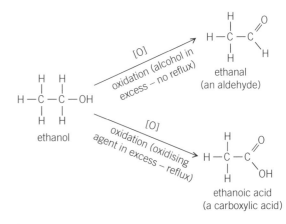

▲ **Figure 2** *Oxidation of a primary alcohol*

Dehydration of alcohols

Alcohols can lose a molecule of water to produce an alkene. This is known as dehydration and is an example of an elimination reaction.

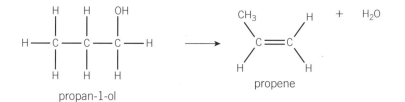

▲ **Figure 3** *Dehydration of an alcohol*

Typical reaction conditions would be using an Al_2O_3 catalyst at 300 °C and 1 atm or refluxing with concentrated sulfuric acid.

Formation of esters

Carboxylic acids react with alcohols in the presence of a strong acid catalyst (concentrated sulfuric acid or concentrated hydrochloric acid) when heated under reflux. This reaction (called esterification) is reversible and reaches equilibrium during refluxing:

alcohol +   Carboxylic acid $\rightleftharpoons$ ester   + water

$CH_3COOH(l) + CH_3CH_2OH(l) \rightleftharpoons CH_3COOCH_2CH_3(l) + H_2O(l)$

This is shown using the skeletal formulae below:

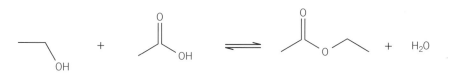

▲ **Figure 4** *Formation of ethyl ethanoate*

Acid anhydrides can also react with alcohols to form esters. Acid anhydrides are more reactive than carboxylic acids, so the reaction goes to completion and does not require a catalyst:

$$(CH_3CO)_2O + CH_3CH_2OH \rightarrow CH_3COOC_2H_5 + CH_3COOH$$
$$\textit{ester}$$

Formation of haloalkanes

Nucleophilic substitution reactions can be used to make haloalkanes from alcohols. The reactants are a halide salt (X^-) in the presence of a strong acid catalyst. The H^+ protonates the oxygen in the alcohol:

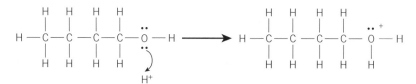

The X^- acts as the nucleophile. The 'leaving group' is a water molecule:

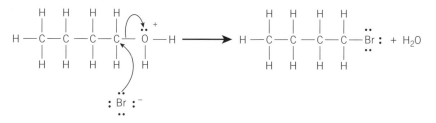

▲ **Figure 5** *Formation of a haloalkane by nucleophilic substitution*

Synoptic link

There is more about carboxylic acids in Topic 13.4, Carboxylic acids and phenols.

Synoptic link

There is more detail on nucleophilic substitution and haloalkanes in Topic 13.2, Haloalkanes.

Summary questions

1 Name the homologous series of the following molecules:

   a butanal

   b propanone

   c propanoic acid

   d butan-2-ol

   e ethyl butanoate.

                (5 marks)

2 Name the two oxidation products that could be obtained from the oxidation of butan-1-ol.

                (2 marks)

3 What colour change would you observe when the following alcohols react with acidified potassium dichromate(VI) solution? Explain your answer.

   a 2-methylpropan-2-ol

   b 2-methylpropan-1-ol.

                (5 marks)

4 What reagents and conditions would you use to turn propan-1-ol into propanal? *(3 marks)*

▲ **Figure 1** *Carboxyl group*

Carboxylic acids contain the **carboxyl** functional group, abbreviated to –COOH. Their systematic names all end with **-oic acid**.

When two carboxyl groups are present, the ending **-dioic acid** is used. Note how the *e* in the name of the alkane is left, for example *ethanedioic* acid.

Reactions of carboxylic acids

Reactions with alkalis

Carboxylic acids react with alkalis to produce salts:

$$CH_3COOH(aq) + NaOH(aq) \rightarrow CH_3COONa(aq) + H_2O(l)$$

Esterification reactions

In the presence of a strong acid, carboxylic acids react with alcohols as follows:

$$Carboxylic\ acid + alcohol \rightarrow ester + water$$

Esters contain an R-COO-R functional group:

▲ **Figure 2** *An ester group*

Tests for carboxylic acids

Although carboxylic acids are weak acids, they will react with carbonates to produce carbon dioxide. Sodium carbonate or sodium hydrogencarbonate solutions are commonly used to test for acids:

$$2CH_3COOH(aq) + Na_2CO_3(aq) \rightarrow 2CH_3COONa(aq) + H_2O(l) + CO_2(g)$$

$$CH_3COOH(aq) + NaHCO_3(aq) \rightarrow CH_3COONa(aq) + H_2O(l) + CO_2(g)$$

The reaction will produce bubbles of carbon dioxide gas, which are readily seen and can be confirmed by testing the gas with limewater, which turns milky.

Phenols

Phenols are compounds that have one or more –OH groups attached directly to a benzene ring:

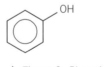

▲ **Figure 3** *Phenol*

Although phenols look similar to alcohols, their chemical reactions are different.

Acidic properties of phenols

Phenols react with alkalis:

$$C_6H_5OH(aq) + NaOH(aq) \rightarrow C_6H_5ONa(aq) + H_2O(l)$$

But they do not react with carbonates.

Test for phenols

When **neutral iron(III) chloride solution** is added to phenol, a **purple** complex is formed. This test allows us to distinguish between phenols and alcohols.

Formation of esters

To produce an ester from a phenol, an acid anhydride is used under alkaline conditions, as phenols do not react with carboxylic acids:

$$C_6H_5OH(aq) + (CH_3CO)_2O \rightarrow CH_3COOC_6H_5 + CH_3COOH$$

Synoptic link

For more details on esterification reactions, see Topic 13.3, Alcohols.

Revision tip

Carboxylic acids are more acidic than phenols.

Revision tip

Carboxylic acids react with alkalis and carbonates.

Phenols react with alkalis, but not carbonates.

Alcohols do not react with alkalis or carbonates.

Revision tip

C_6H_5ONa is the salt, sodium phenoxide. It is an ionic compound: $C_6H_5O^- Na^+$.

Revision tip

Remember that alcohols can form esters from carboxylic acids or acid anhydrides. With a phenol, you must use an acid anhydride.

Summary questions

1   Name the following carboxylic acids:
    a $CH_3CH_2CH_2CH_2CH_2COOH$
    b $HOOCCH_2CH_2CH_2COOH$
    c $CH_3CH(CH_3)CH_2COOH$. *(3 marks)*

2   Write the equation for the reaction of propanoic acid with potassium hydroxide. *(2 marks)*

3   State the names and formulae of the reactants needed to make propyl butanoate. *(4 marks)*

4   A student had unlabelled samples of propan-1-ol, 4-methylphenol, and ethanoic acid. Which of the samples would:
    a   give a purple colour with neutral iron(III) chloride
    b   react with sodium carbonate solution, producing carbon dioxide gas
    c   react with acidified potassium dichromate(VI)
    d   react with sodium hydroxide solution? *(4 marks)*

5   Write an equation for the reaction of phenol with sodium carbonate solution. *(1 mark)*

Heating under reflux

This technique is used to heat volatile reactants together in order to allow a chemical reaction to take place without any of the mixture escaping. This increases yield and reduces the risk of fires.

The reactants and products evaporate on heating but change back into liquids in the condenser and fall back into the reaction mixture in the flask.

1   Put the reactants in a pear-shaped or round-bottomed flask. Add a few anti-bumping granules. Attach a condenser vertically to the flask as shown in the diagram. Do NOT stopper the condenser.
2   Connect the condenser to the water supply.
3   Heat so that the liquid boils gently, using a Bunsen burner or heating mantle.

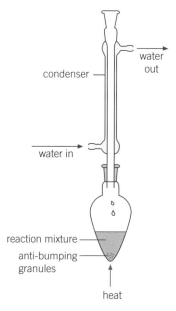

▲ **Figure 1** *Heating under reflux*

Simple distillation

This technique is used to separate two or more miscible liquids with different boiling points. For details of the apparatus used see Figure 2. When the mixture is heated the liquid with the lowest boiling point evaporates first.

The mixture is heated until the thermometer shows the vapour temperature to be at the boiling point of the desired liquid. A clean receiver beaker is placed below the condenser to collect the condensed liquid. When the temperature on the thermometer rises above the boiling point of the desired liquid, the distillation process is stopped.

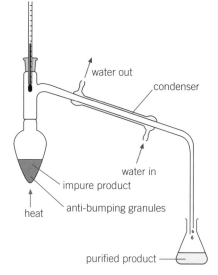

▲ **Figure 2** *Simple distillation*

1 What is the correct name for $CH_3CH_2CH_2CH_2CH(CH_3)CH_3$?

  A heptane

  B 5-methylhexane

  C 2-methylhexane

  D 1-methylheptane   *(1 mark)*

2 Which of these homologous series does not contain oxygen?

  A carboxylic acid

  B phenol

  C ketone

  D alkane   *(1 mark)*

3 $CH_3C(CH_3)(OH)CH_3$ is

  A a tertiary alcohol

  B an aldehyde

  C a carboxylic acid

  D a ketone   *(1 mark)*

4 Which of these reactions do primary alcohols undergo?

  1 Nucleophilic substitution   **2** Oxidation   **3** Dehydration

  A 1 only

  B 1 and 2 only

  C 2 and 3 only

  D 1, 2, and 3   *(1 mark)*

5 Which of the following is the formula of an ester?

  A C_6H_5COOH

  B $C_6H_5COOC_2H_5$

  C $C_6H_5COOCOC_6H_5$

  D $C_6H_5CH_2OH$ *(1 mark)*

6 Which statement correctly describes the acid-base properties of carboxylic acids and phenols?

  A Carboxylic acids react with sodium carbonate but not sodium hydroxide.

  B Carboxylic acids react with sodium hydroxide but not sodium carbonate.

  C Phenols react with sodium carbonate but not sodium hydroxide.

  D Phenols react with sodium hydroxide but not sodium carbonate.

                                                                  (1 mark)

7 What are the correct conditions for converting C_2H_5Br into C_2H_5OH?

  A Heat under reflux with NaOH(aq) with ethanol as solvent.

  B Heat under reflux with acidified dichromate(VI).

  C Distil with acidified dichromate(VI).

  D React with a carboxylic acid in the presence of a strong acid catalyst.

                                                                  (1 mark)

8 Which of the following haloalkanes has the highest boiling point?

  A CH_3F

  B C_2H_5F

  C C_2H_5Cl

  D $C_2H_4Cl_2$ *(1 mark)*

9 A student has an unlabelled sample of a substance which is suspected to be a phenol.

The student adds a range of reagents to the separate samples of the substance.

State what the student would expect to see in each case if the substance is a phenol.

   a Adding neutral iron(III) chloride. *(1 mark)*

   b Adding sodium hydrogen carbonate. *(1 mark)*

   c Esterifying the sample with a carboxylic acid and a strong acid catalyst. *(1 mark)*

10 A substance X has the molecular formula $C_4H_{10}O$. When heated under reflux with acidified sodium dichromate(VI) there is no colour change.

   a State the colour of the reaction mixture. *(1 mark)*

   b Explain what this result tells you about substance X. *(2 marks)*

   c Give the structure and name of substance X. *(2 marks)*

   d There are two isomers of X that contain an –OH group at the end of the carbon chain. Give the structure of these isomers. *(2 marks)*

   e Give the products that would be obtained when these isomers are separately oxidised with acidified sodium dichromate(VI). *(4 marks)*

11 The physical properties and reactions of haloalkanes depend on the fact that the C–Hal bond is polar.

   a Describe and explain the polarity of the C–Hal bond. *(2 marks)*

   b Give an example of a physical property that depends on the polarity of the C–Hal bond. *(1 mark)*

   c Nucleophiles are attracted to the polar bond and can substitute the halogen atom. Describe the mechanism of this reaction. *(3 marks)*

14.1 Risk and benefits of chlorine

Specification reference: ES(n), WM(g)

Greener industry

Chemical manufacturers pay careful attention to the area of green chemistry, which involves several different considerations.

Batch processes and continuous processes

During a batch process, products are removed at the end of the reaction and the vessel is cleaned for the next batch. This is cost-effective for small quantities but is time-consuming and may require a larger workforce.

In a continuous process, the starting materials are fed in at one end of the plant and the product emerges at the other end. This is more easily automated, operating for months at a time, but is usually designed to make one product.

Raw materials

Raw materials are usually obtained from the ground or the atmosphere. They are converted into feedstocks – the reactants which are fed in at the start of the process.

Costs and efficiency

The efficiency of a chemical process depends on many factors. High temperatures increase the rate of a reaction, but may affect the position of equilibrium and reduce yields. High pressure can improve the yield, but requires more expensive reaction vessels and has safety implications. Recycling unreacted feedstock is an important way of reducing costs. Efficient use of energy is also essential.

Plant location

Chemical plants were traditionally sited near sources of raw materials. Nowadays, factors which affect the location of a chemical plant include availability of good transport links, skilled labour, cheap energy, and plentiful water.

Health and safety

Safety is a major consideration, and companies have to abide by legislation. Every stage of a manufacturing process is checked to reduce exposure of employees to hazardous chemicals or procedures.

Waste disposal

There are strict limits on the amount of hazardous chemicals that can be released. All chemical waste must be treated before disposal.

Revision tip

Most bulk chemicals are made by continuous processes.

Revision tip

Raw materials can be obtained from rocks, crude oil, seawater, or the atmosphere.

Key term

Feedstock: Feedstocks are chemicals produced from raw materials. They are the reactants in the chemical reactions in the process.

Synoptic link

You learned about rate of reaction in Topic 10.1, Factors affecting reaction rates, and about equilibrium in Topic 7.1, Chemical equilibrium.

Revision tip

You should link your knowledge of green chemistry to any industrial process you study.

Synoptic link

You learned about the halogens in Topic 11.3, The p-block: Group 7.

Chlorine

Chlorine is produced by electrolysis of the feedstock brine, from the raw material rock salt. This is a continuous process and requires large quantities of electricity.

Chlorine is transported under pressure as a liquid in special containers. They are lined with steel and designed to vent small quantities of chlorine if the pressure or temperature gets too high, rather than risk a catastrophic explosion.

These precautions are necessary, as chlorine is a toxic gas. Workers using chlorine must follow strict regulations. About 50 million tonnes of chlorine are used every year in water purification and the production of bleach.

Summary questions

1 Explain the difference between a batch process and a continuous process. *(1 mark)*

2 Suggest if the following processes are batch or continuous:
   a catalytic cracking of gas oil in the petrol industry
   b fermentation of grapes to make wine
   c the manufacture of ammonia in the Haber process. *(3 marks)*

3 Suggest why locating a new chemical plant on an existing site of chemical manufacture can lower costs. *(3 marks)*

14.2 Atom economy

Specification reference: ES(a)

Percentage yield and atom economy

The percentage yield for a reaction is calculated using the equation:

$$\% \text{ yield} = \frac{\text{actual mass of product}}{\text{theoretical maximum mass of product}} \times 100$$

The atom economy for a reaction is calculated using the equation:

$$\% \text{ atom economy} = \frac{\text{relative formula mass of useful product}}{\text{relative formula mass of reactants used}} \times 100$$

> ### 🖩 Worked example: Calculating percentage yield and atom economy
>
> Ethanol can be converted into ethene in the following reaction:
>
> $$C_2H_5OH \rightarrow C_2H_4 + H_2O$$
>
> In one reaction, 23.0 g of ethanol produced 6.00 g of ethene. Calculate the percentage yield and the atom economy of this reaction.
>
> **Step 1:** moles of ethanol used $= \dfrac{\text{mass}}{M_r} = \dfrac{23.0}{46.0} = 0.50$ mol
>
> **Step 2:** We can see from the equation that 0.50 mole of ethanol should, in theory, produce 0.50 mole of ethene, or 14.0 g (28.0×0.50).
>
> So $\%$ yield $= \dfrac{6.0}{14.0} = 43\%$
>
> **Step 3:** $\%$ atom economy $= \dfrac{28.0}{46.0} = 61.0\%$

Percentage yield used to be a significant factor in deciding if a chemical process was economically viable. With green chemistry high on the agenda, atom economy is just as important.

Atom economy is a way of measuring the efficient use of atoms in a chemical reaction. The higher the atom economy, the more atoms from the reactant molecules end up in the useful product, and the fewer end up in waste products. If the atom economy is 100% there are no waste products, and this has environmental benefits.

> ### Revision tip
> Add up all formula masses of all the reactants when calculating % atom economy.

> ### Revision tip
> Learn the equations for % yield and % atom economy.

> ### Synoptic link
> The theoretical yield is calculated using your knowledge of stoichiometric relationships, which you covered in Topic 1.3, Using equations to work out reacting masses.

> ### Synoptic link
> You learned how to calculate formula masses in Topic 1.1, Amount of substance.

> ### Revision tip
> If there is only one product, the % atom economy is 100%, as there are no waste products.

Summary questions

1 Epoxyethane, C_2H_4O is an important feedstock in the chemical industry. It is made by the following reaction:

$$2C_2H_4 + O_2 \rightarrow 2C_2H_4O$$

  a Calculate the atom economy of the reaction. (*1 mark*)
  b Suggest what happens to the unreacted ethene in the reaction. (*1 mark*)

2 Place these types of chemical reaction in order from highest to lowest atom economy. Explain your answer. (*2 marks*)
   • addition
   • elimination
   • substitution.

3 Chloroethene, C_2H_3Cl is the monomer for poly(chloroethene) manufacture. It can be produced in a two-step process:
   Step 1: $C_2H_4 + Cl_2 \rightarrow C_2H_4Cl_2$
   Step 2: $C_2H_4Cl_2 \rightarrow C_2H_3Cl + HCl$
   Calculate the atom economies for step 1 and step 2. (*2 marks*)

1 Which of the following chemicals are produced by batch processes?

   A Ammonia in the Haber Process

   B Ethanol by reacting ethene and steam

   C Pharmaceuticals to be tested for biological effect

   D Sodium hydroxide by electrolysis of brine.   *(1 mark)*

2 Which of the following considerations are taken into account when deciding the location of a chemical plant?

    1 Availability of labour

    2 Transport links

    3 Availability of raw materials.

   A 1 only   **B** 2 only

   C 2 and 3 only   **D** 1, 2, and 3   *(1 mark)*

3 What is the correct formula for calculating atom economy?

   A $\dfrac{\text{relative formula mass of useful product}}{\text{relative formula mass of reactants used}} \times 100$

   B $\dfrac{\text{relative formula mass of reactants used}}{\text{relative formula mass of useful product}} \times 100$

   C $\dfrac{\text{actual mass of product}}{\text{theoretical maximum mass of product}} \times 100$

   D $\dfrac{\text{theoretical maximum mass of product}}{\text{actual mass of product}} \times 100$ *(1 mark)*

4 The equation for the formation of ethene from ethanol is $C_2H_5OH \rightarrow C_2H_4 + H_2O$. What is the atom economy of this reaction?

   A 1.64%   **B** 28%

   C 39%   **D** 61%   *(1 mark)*

5 A student carried out a reaction to produce propanone from propan-2-ol:

$$CH_3CH(OH)CH_3 + [O] \rightarrow CH_3COCH_3 + H_2O$$

[O] represents an oxygen from the oxidising agent.

   a What reagent and conditions would the student use for this reaction?   *(3 marks)*

   b What functional groups do propanone and propan-2-ol contain? *(2 marks)*

   c Calculate the atom economy of the reaction.   *(1 mark)*

   d The student started with 6.0 g of propan-2-ol. Calculate the amount in moles that the student used.   *(1 mark)*

   e What amount (in moles) of propanone would the student expect to make?   *(1 mark)*

   f What is the theoretical maximum mass of propanone in this reaction?   *(1 mark)*

   g The student's actual yield was 2.61 g. Calculate the percentage yield.   *(1 mark)*

6 A student carried out a reaction to produce ethanol from bromoethane:

$$C_2H_5Br + H_2O \rightarrow C_2H_5OH + HBr$$

   a Calculate the atom economy of the reaction.   *(1 mark)*

   The student reacted 5.445 g of C_2H_5Br and produced 1.426 g of C_2H_5OH.

   b Calculate the percentage yield.   *(4 marks)*

15.1 Atmospheric pollutants

Specification reference: DF(k)

Production and effects of pollutants from petrol

Combustion of hydrocarbons in car engines results in a wide range of pollutants being produced. These atmospheric pollutants can have significant environmental implications.

Particulates

Particulates are small carbon particles below 2.5×10^{-12} m formed from burning fossil fuels. They can penetrate into the human body causing lung cancer and heart attacks. They also make surfaces dirty.

Unburnt hydrocarbons

These are formed from incomplete combustion or leakage from fuel tanks. They can go on to form photochemical smog.

Carbon monoxide

CO is formed from incomplete combustion of hydrocarbons. It is a toxic gas which is directly harmful to humans. Catalytic converters can oxidise CO into CO_2.

Revision tip

Particulates and carbon monoxide harm humans directly.

Carbon dioxide

CO_2 is formed from the complete combustion of hydrocarbons. It is a greenhouse gas and is responsible for global warming.

Nitrogen oxides

Nitrogen monoxide is formed from the reaction of oxygen and nitrogen (from the air) at high temperatures inside car engines. Nitrogen monoxide is readily oxidised to nitrogen dioxide by oxygen in the atmosphere. Nitrogen oxides (often written as NO_x) dissolve in rainwater to cause acid rain. Catalytic converters can reduce nitrogen monoxide to nitrogen, N_2.

Revision tip

Exposure to NO_2 has been linked to respiratory illnesses such as asthma or bronchitis.

Sulfur oxides

Sulfur oxides, SO_x, are formed from the oxidation of sulfur impurities in fuels. Sulfur oxides form acid rain. It is possible to remove the sulfur from fuel before burning – this is called desulfurisation.

Revision tip

Global warming and acid rain harm humans indirectly.

Summary questions

1 Write an equation for the formation of nitrogen monoxide in a car engine. Describe one polluting effect of nitrogen monoxide. *(2 marks)*

2 Describe the harmful effects of carbon monoxide. Suggest an equation for the formation of carbon monoxide from methane. *(2 marks)*

3 A lean burn engine uses a higher ratio of air to petrol vapour than other engines. Explain why less unburnt hydrocarbons are produced. *(1 mark)*

Key terms

Carbon neutral: On burning, a carbon-neutral fuel only releases as much CO_2 as the plant it came from absorbed when it was growing.

Sustainable: Able to be used without negative impact on future generations.

Renewable: Renewable fuels will not run out, because they come from sources such as plants, which can be regrown.

Finite resource: A finite resource will run out. Crude oil is a finite resource.

Revision tip

A fuel can only be considered truly carbon neutral if its production, transportation, and so on are carbon neutral as well.

Alternative fuels for a car engine

People need to make choices, both now and in the future, regarding the use of fuels. We need to weigh up the risks, benefits, and sustainability for each fuel.

A variety of fuel choices for the future are discussed in the table below.

Alternative	Is it sustainable?	Benefits	Risks
Diesel	No; crude oil is running out	Less CO_2 produced than from a petrol engine; already sold at petrol stations	Produces more NO_x and particulates than a petrol engine; particulates can irritate lungs
LPG or autogas	No; crude oil is running out	Less CO, CO_2, C_xH_y, and NO_x than from a petrol engine; petrol engines easily converted	Needs to be stored under pressure, so that it is a liquid
Ethanol	Possibly not; large amounts of energy needed for cultivating sugar cane for fermentation	Less CO, SO_2, and NO_x than from a car engine; ethanol has a high octane number; sugar cane absorbs CO_2 in growth	Highly flammable
Biodiesel	Can be made from waste plant and animal oils and fats, so renewable, but fossil fuels may be used as an energy source in production	Living things have absorbed CO_2; it is biodegradable; less CO, C_xH_y, SO_2, and particulates than from a diesel engine	NO_x emissions higher than a diesel engine
Hydrogen	Only if the electricity needed for electrolysis of water is from a renewable source such as solar cells	Water is the only product of combustion	Highly flammable; high-pressure fuel tank needed to store it as a liquid

Summary questions

1   What problems may arise in the storage of hydrogen in a fuel tank? *(1 mark)*

2   Biodiesel can be manufactured from soya beans. Give one possible advantage and one disadvantage of such a fuel. *(2 marks)*

3   Suggest why ethanol is only likely to be a suitable choice in certain parts of the world. *(1 mark)*

1 Which of the following pollutants is produced during complete combustion of hydrocarbons?

   A carbon monoxide   **B** particulates

   C carbon dioxide   **D** unburnt hydrocarbons   *(1 mark)*

2 Which of the following statements best explains why nitrogen oxides are formed in internal combustion engines?

   A Nitrogen from the fuel reacts with oxygen from the air.

   B Nitrogen from the air reacts with oxygen from the fuel.

   C Nitrogen and oxygen from the fuel react together.

   D Nitrogen and oxygen from the air react together.   *(1 mark)*

3 Which of the following statements best explains why sulfur oxides are formed in internal combustion engines?

   A Sulfur from the fuel reacts with oxygen from the air.

   B Sulfur from the air reacts with oxygen from the fuel.

   C Sulfur and oxygen from the fuel react together.

   D Sulfur and oxygen from the air react together.   *(1 mark)*

4 What is an advantage of hydrogen fuel?

   A It is produced by electrolysis.

   B Water is the only product of combustion.

   C It is carbon neutral.

   D It can be easily stored.   *(1 mark)*

5 Why do some fuels produce less nitrogen oxide pollution than petrol?

   A They do not contain nitrogen atoms in their molecules.

   B They are more likely to undergo complete combustion.

   C They have a greater energy density.

   D They burn at lower temperatures.   *(1 mark)*

6 What is the best definition of a *sustainable* fuel?

   A It will never run out.

   B It can be used without negative impact on future generations.

   C It lasts a long time.

   D It does not contribute to global warming.   *(1 mark)*

7 Write equations for the following reactions which produce different pollutants.

   a The complete combustion of decane, $C_{10}H_{22}$.   *(1 mark)*

$$C_{10}H_{22} + \tfrac{31}{2}O_2 \rightarrow 10CO_2 + 11H_2O$$

   b The oxidation of nitrogen monoxide to nitrogen dioxide by oxygen in the air.   *(1 mark)*

$$NO + \tfrac{1}{2}O_2 \rightarrow NO_2$$

   c The oxidation of sulfur dioxide to sulfur trioxide by oxygen in the air.   *(1 mark)*

$$SO_2 + \tfrac{1}{2}O_2 \rightarrow SO_3$$

   d The formation of acid rain from sulfur trioxide.   *(1 mark)*

$$SO_3 + H_2O \rightarrow H_2SO_4$$

8 Explain why biodiesel is considered a sustainable fuel.   *(4 marks)*

1.1

1 a 342

b 128   *[1 for each]*

2 92.3%   *[1]*

3 Relative isotopic mass is the relative mass of a single isotope *[1]*; relative atomic mass may be the (weighted) average mass of several isotopes *[1]*

4 C and O have different masses *[1]*. An atom of C has a smaller mass than an atom of O *[1]*.

5 a 79.99   *[1]*

b 20% ^{11}B and 80% ^{10}B *[1]*

6 % of O = 16.33% *[1]*. Empirical formula of the compound = $C_6H_{10}O$ *[1]*. 2 empirical formula units are needed to make a M_r of 196, so molecular formula = $C_{12}H_{20}O_2$ *[1]*.

1.2

1 a S^{2-}

b Rb^+ *[1 mark each]*

2 a $Ba(OH)_2$

b Al_2O_3

c Na_2SO_4 *[1 mark each]*

3 a $CaO(s) + 2HCl(aq) \rightarrow CaCl_2(aq) + H_2O(l)$. *[1 mark formulae, 1 mark balancing, 1 mark state symbols]*

b $Ba^{2+}(aq) + SO_4^{2-}(aq) \rightarrow BaSO_4(s)$ *[1 mark formulae, 1 mark state symbols]*

1.3

1 36.7 g   *[1]*

2 M_r of UF_6 = 352.1, M_r of ClF_3 = 109 *[1]*. Moles of UF_6 formed = 100/352.1 = 0.284 mol, so moles of ClF_3 needed = 0.568 *[1]*. Mass of ClF_3 needed = 0.568 × 109 = 61.9 g *[1]*.

3 Moles U = 1000/238.1 = 4.20 mol; mol ClF_3 = 1000/109 = 9.17 *[1]*. ClF_3 is in excess and U is limiting reagent *[1]*. Expected yield of UF_6 = 352.1 × 4.20 = 1478 g *[1]* % yield = (1.36/1.478) × 100 = 92.0% *[1]*

1.4

Challenge: dilution factor = $\dfrac{250}{(3.57)}$ = 70.0. Original concentration = 70.0 × 0.045 = 3.15 mol dm^{-3}

1 a 0.800 mol dm^{-3}

b 0.040 mol dm^{-3}

c 0.125 mol dm^{-3} *[1 mark each]*

2 Moles of HCl = 28.6/36.5 = 0.7836 mol *[1]*. Concentration = 0.7836/0.250 = 3.13 mol dm^{-3} *[1]*. Dilution factor = 10/250, so concentration of diluted solution = (10/250) × 3.13 = 0.125 mol dm^{-3} *[1]*

3 Moles H_2SO_4 = 2.09 × 10^{-4} mol *[1]*. Moles KOH = 4.18 × 10^{-4} mol *[1]*. Concentration = $\dfrac{4.18 \times 10^{-4}}{0.0200}$ = 0.0209 mol dm^{-3} *[1]*. M_r of KOH = 56, so concentration = 0.0209 × 56 = 1.17 g dm^{-3} *[1]*.

1.5

1 a 0.00182%

b 400 ppm [1 mark each]

2 M_r (H_2O_2) = 34, so 2.40 g H_2O_2 = 0.07059 mol *[1]*. Moles of O_2 formed = 0.07059/2 = 0.03429 mol *[1]*. Volume of O_2 at RTP = 0.03429 × 24.0 = 0.847 dm^3 *[1]*

3 $n = \dfrac{pV}{RT}$ *[1]*. 360 dm^3 = 0.36 m^3 *[1]*, so $n = \dfrac{(200\,000 \times 0.36)}{(8,314 \times 400)}$ = 21.65 mol *[1]*.

2.1

1 It was realised that the atom was able to be divided into smaller particles OR consisted of positive and negative parts.   *[1]*

2 a 1 proton + 2 neutrons

b 20 protons and 27 neutrons

c 11 protons + 12 neutrons   *[1 mark for each part question]*

3 The atom is mostly empty space (so most passed through); the atom contains a positive, dense / small nucleus (so some particles bounce back).   *[2]*

2.2

1 To overcome the repulsion *[1]* between positive nuclei *[1]*

2 a ^1_0n *[1]*

b $^{16}_8O$ *[1]*

3 a $^3_2He \rightarrow {}^4_2He + 2{}^1_1H$ *[1]*

2.3

1

↑↓	↑	↑

4 electrons in total arranged as above.   *[2]*

2 a $1s^2\ 2s^2\ 2p^6\ 3s^2\ 3p^1$ *[1]*

b $1s^2\ 2s^2\ 2p^6$ *[1]*

c $1s^2 2s^2 2p^6 3s^2 3p^6$ *[1]*

3 a $4s^2\ 3d^3$ (or reversed)

b $6p^4$ *[1 mark each*

3.1

1 a ionic   *[1]*

b covalent   *[1]*

2 $\left[Mg \right]^{2+}$ $\left[\begin{smallmatrix} \times\times \\ \times\ O\ \times \\ \bullet\ \times \end{smallmatrix} \right]^{2-}$

[1 mark: correct structures of Mg and O, 1 mark correct charges on ions]

3

[1 mark: correct number of electrons on nitrogen atoms (5) and oxygen atom (6); 1 mark: triple bond between N and N, single bond between N and O; 1 mark: single bond shown as a dative bond.]

3.2

1 a octahedral, 90°   [2]

  b trigonal planar, 120° *[1 mark for shape, 1 mark for bond angle in each case]*

2 3 bonding pairs and 1 lone pair around P atom *[1]*. 3 lone pairs around each. Cl *[1]*.

There are 4 pairs of electrons around the central P atom *[1]*. These repel and get as far apart as possible (to minimise repulsion) *[1]*. Electrons arranged tetrahedrally *[1]*. One lone pair and 3 bonding pairs so shape of molecule is pyramidal *[1]*. Extra repulsion from lone pair reduces tetrahedral bond angle (109.5°) to 107° *[1]*.

3 a 2 double bonds from S to O atoms *[1]*. 2 single bonds from S to O atoms *[1]*. Extra electron on the single-bonded O atoms shown clearly *[1]*.

  b

tetrahedral
SO_4^{2-} ion

Shape [1], bond angle. [1].

4.1

1 a i Negative   **ii** Temperature increases *[2]*

  b Energy is released by forming bonds AND energy is required to break bonds *[1]* Energy released is greater than energy required *[1]*

2 Energy released to the water = 25 707 J = 25.707 kJ *[1]*. Moles of methanol burnt = (1.60)/32 = 0.0500 mol *[1]*. Energy released by 1 mole = (25.707)/(0.05) = 514.14 *[1]*. $\Delta H = -514$ kJ mol^{-1} (3 s.f.) *[1]*

3 Moles of NI$_3$ = 1000/394.7 = 2.524 mol *[1]*. Ratio of moles reacted: moles in equation = 2.524/2 = 1.267 *[1]*. $\Delta H = -290 \times 1.267 = -367$ kJ, so 367 kJ of energy is released *[1]*.

4.2

1 $\Delta_f H$ [C$_2$H$_4$], $\Delta_f H$ [C$_2$H$_6$] [1 mark each]. Any additional substance, e.g. H$_2$ loses one of these 2 marks.

2 $\Delta_f H$(products) $-\Delta_f H$ (reactants) *[1]* (or correct use of the enthalpy changes in an enthalpy cycle); $\Delta_f H$(products) = $2 \times -285.8 = -571.6$ *[1]*. $\Delta_f H$ (reactants) = $+50.6$ *[1]* $\Delta_r H = -571.6 - 50.6 = -622.2$ kJ mol^{-1}. *[1]*

3 a Top line of cycle shown as 3C(s) + 4H$_2$(g) → C$_3$H$_8$(g) *[1]* Combustion products at bottom = 3CO$_2$(g) + 4H$_2$O(l) *[1]* + 5O$_2$ included on each downward arrow. *[1]*

  b $\Delta H_2 = (3 \times -394) + (4 \times -286) = -2326$ kJ mol^{-1} *[1]*
$\Delta H_3 = -2219$ kJ mol^{-1} *[1]*. ΔH_1 ($\Delta_f H$ [C$_3$H$_8$]) = $-2316 - (-2219) = -97$ kJ mol^{-1}. *[1]*

4.3

1 a C=C is stronger than C–C *[1]*

  b C–C is longer than C=C. *[1]*

2 a Bonds broken = 4 C–H + 2O=O AND bonds formed = 2C=O + 4 O–H *[1]*. Energy changes = 2468 AND −3466 *[1]*. Enthalpy change of reaction = −818 kJ mol^{-1} *[1]*.

  b Bond enthalpies are average values OR H$_2$O is a liquid in its standard state, bond enthalpies are for gaseous molecules *[1]*.

3 C$_3$H$_6$ + 4$\frac{1}{2}$O$_2$ → 3CO$_2$ + 3H$_2$O *[1]*
Bonds formed = 6C=O + 6O–H = 7614 *[1]*
Bonds broken = 3C–C + 6C–H + 4$\frac{1}{2}$ O=O = 3C–C + 4719 *[1]*
$\Delta H = -2091 = (3C–C + 4719) - 7614 = 3C–C - 2895$ *[1]*
3C–C = $-2091 + 2895 = 804$. *[1]*
C–C = $\frac{804}{3}$ = + 268 kJ mol^{-1} *[1]*
Significantly lower than usual C–C bond enthalpy so bond is easier to break in cyclopropane. *[1]*

5.1

1 a Add silver nitrate solution, *[1]* a cream precipitate shows presence of bromide ions. *[1]*

  b Add barium chloride solution *[1]*, a white precipitate shows presence of sulfate ions. *[1]*

2 a blue

  b white

  c brown

  d yellow   [8]

3 a iron(II) sulfate.   [2]

  b Fe^{2+}(aq) + 2OH$^-$(aq) → Fe(OH)$_2$(s) *[1 formulae, 1 balancing and state symbols]* Ba^{2+}(aq) + SO$_4^{2-}$ (aq) → BaSO$_4$(s).   [2]

5.2

Challenge:

Octahedral

Challenge:

Group 1 ions have a small (1+) charge and have a relatively large ionic radius (compared to other ions in the same period). These two factors reduce the electrostatic attraction between the ions and the delocalised electrons.

1 Calcium ions *[1]* are arranged in a regular pattern / in layers *[1]* to form a giant 3-dimensional lattice *[1]* delocalised electrons surround the ions. *[1]*

2 Calcium oxide: high melting (and boiling) point *[1]*, probably soluble in water *[1]* conducts electricity when molten or in solution. *[1]*

3 Silicon dioxide has a covalent network / giant covalent structure *[1]* carbon dioxide has a covalent molecular structure *[1]* weak intermolecular bonds between carbon dioxide molecules *[1]* (no molecules in silicon dioxide so) all bonds are (strong) covalent bonds *[1]* (much) more energy is needed to break covalent bonds than intermolecular bonds. *[1]*

5.3

1 C–Cl: C δ^+, Cl δ^- *[1]*   Cl–F: Cl δ^+, F δ^- *[1]*

   H–N: H δ^+, N δ^- *[1]*   Cl–S: Cl δ^-, S δ^+. *[1]*

2 a instantaneous dipole–induced dipole *[1]*

   b Electrons in a molecule are in continuous random motion *[1]*. At a particular instant, the electrons may be distributed unevenly *[1]*. This creates an instantaneous dipole. *[1]* The dipole induces a dipole on a neighbouring molecule, creating an induced dipole *[1]*. There is an electrostatic attraction between the two dipoles *[1]*.

3 CCl_2F_2 is tetrahedral *[1]* C–Cl and C–F bonds are polar but C–F bonds are more polar (or can be shown on diagram using appropriate δ^- δ^+ convention) *[1]*; centre of positive charge is not in same place as centre of negative charge / dipole do not cancel out / charges are not arranged symmetrically *[1]*; there is an overall dipole. *[1]*

5.4

1 N is a small atom with a δ^- charge / small electronegative atom *[1]*. H has a δ^+ charge (because it is bonded to a N atom) *[1]* there is a lone pair of electrons on the N atom. *[1]*

2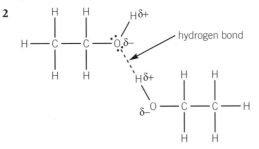

   Correct charges on O and H atoms *[1]* Lone pair on O *[1]* shown pointing to H *[1]* H....O–H shown in a line / 180° bond angle around H atom. *[1]*

3 id–id bonds between methane molecules, H-bonds (and id–id) between water and between hydrogen fluoride *[1]*. H-bonds are stronger than id–id *[1]*. Water has more H-bonds per molecule than hydrogen fluoride *[1]*, stronger and/or greater number of bonds = more energy needed to separate molecules *[1]*.

6.1

Challenge:

Lines get closer at high frequency: High frequency lines arise from transitions with large ΔE values; these are produced from transitions from higher energy levels; high energy levels are much closer together than lower energy levels; transitions from adjacent energy levels

will have very similar ΔE values hence produce light of similar frequencies.

Several series of lines: Lines are caused when electrons drop to a lower energy level; all transitions ending at, e.g. $n = 1$ produce one series; other series are produced by drops to $n = 2$, $n = 3$, etc.

1 a yellow *[1]*

   b (pale) green *[1]*

2 Differences in appearance: emission spectra consists of coloured / bright lines on a dark background *[1]* absorption spectra consists of black lines on a coloured background *[1]*. Difference in formation: emission spectra formed by electron falling from higher to lower energy levels and emitting light *[1]* absorption spectra formed by electrons being excited from lower to higher energy levels and absorbing light. *[1]*

3 Electrons drop from higher to lower energy levels *[1]* emit energy in the form of light *[1]* energy lost is related to frequency of photon / $\Delta E = h\nu$ *[1]* each drop in energy levels causes one line in the spectrum *[1]* series of lines caused when electrons drop to a specific level, e.g. $n = 2$. *[1]*

6.2

1 a Infrared *[1]*

   b Visible light (accept ultraviolet) *[1]*

2 a Visible has higher energy / greater frequency / shorter wavelength (any two) *[2]*

   b visible light excites electrons to higher energy levels OR can cause bond breaking *[1]* infrared causes bonds to vibrate (with greater energy) *[2]*

3 Energy of photon absorbed when bond breaks = 7.227 $\times 10^{-19}$ J *[1]*. Energy needed to break 1 mole of bonds = $7.227 \times 10^{-19} \times 6.02 \times 10^{23} = \times 10^3$ J *[1]*. 435 kJ mol^{-1}. *[1]*

6.3

Challenge:

Reaction of methane with chlorine:

$$CH_4 + Cl \rightarrow CH_3 + HCl$$
$$CH_3 + Cl_2 \rightarrow CH_3Cl + Cl$$
$$\text{overall: } CH_4 + Cl_2 \rightarrow CH_3Cl + HCl$$

Depletion of ozone:

$$O_3 + Cl \rightarrow ClO + O_2$$
$$ClO + O \rightarrow Cl + O_2$$

1 Homolytic fission: one electron goes to each atom; heterolytic: both electrons go to a single atom. *[1]*

   Homolytic: radicals are formed; heterolytic: ions are formed. *[1]*

2 a Ozone is broken down to form oxygen molecule and an oxygen atom / $O_3 + h\nu \rightarrow O_2 + O$. *[1]*

   b Prevents high frequency ultraviolet radiation reaching the Earth's surface *[1]* high frequency uv can cause skin cancer / eye cataracts. *[1]*

3 $NO + O_3 \rightarrow NO_2 + O_2$ *[1]*

$NO_2 + O \rightarrow NO + O_2$ *[1]*

6.4

1 The bonds vibrate *[1]* with greater energy / more *[1]*

2 2500–3200: O–H bond *[1]* in a carboxylic acid *[1]*; 1715: C=O bond *[1]* in a carboxylic acid *[1]*

3 $\lambda = \dfrac{c}{f}$ *[1]* $= \dfrac{3.00 \times 10^8}{9.04 \times 10^{13}} = 3.318 \times 10^{-6}\,m$ *[1]* $= 3.318 \times 10^{-4}\,cm$, so wavenumber $= \dfrac{1}{3.318} \times 10^{-4} = 3013\,cm^{-1}$ *[1]*

6.5

Challenge:

a 3 peaks are: $^{35}Cl^{35}Cl$, $^{35}Cl^{37}Cl$ and $^{37}Cl^{37}Cl$; m/z values are 70, 72, and 74.

b Abundance of $^{35}Cl^{35}Cl = (75.78/100)2 \times 100 = 57.4\%$, $^{37}Cl^{37}Cl = (24.22/100)2 \times 100 = 6.87\%$, abundance of $^{35}Cl^{37}Cl = 2 \times (75.78/100) \times (24.22/100) \times 100 = 36.7\%$ (or by subtraction from previous 2 answers).

1 Number of different isotopes, mass number / isotopic mass of each isotope, relative abundance of each isotope. *[1 mark each]*

2 $m/z\ 78$ = molecular ion / M^+ ion / unfragmented molecule with a 1^+ charge *[1]* $m/z = 79$ is M^+ ion containing a ^{13}C atom. *[1]*

7.1

1 When the concentrations of the species in the reaction are constant and the rate of the forward reaction and backward reaction are equal. *[1]*

2 If the reaction starts with the reactants from the left-hand side the rate of the forward reaction is initially greater than the rate of the reverse reaction *[1]*. If the reaction starts with the substances on the right-hand side of the equation the rate of the reverse reaction is initially greater. At equilibrium the rate of both reactions is equal. *[1]*

3 Increasing the temperature means more of the particles will be in the gaseous state *[1]*. The position of equilibrium will be further to the right. *[1]*

7.2

1 a $K_c = \dfrac{[SO_3]^2}{[SO_2]^2\,[O_2]}$ *[1]*

b $K_c = \dfrac{[NO_2]^2}{[N_2O_4]}$ *[1]*

2 a $K_c = \dfrac{[C_2H_5OH]}{[C_2H_4]\,[H_2O]}$ *[1]*

b The position of equilibrium would move to the side with fewer moles – to the right. K_c would not change. *[1]*

c The position of equilibrium would move to the endothermic side – to the left. K_c would decrease. *[1]*

d Equilibrium is reached more quickly (but the position of equilibrium is not affected). *[1]*

7.3

1 $K_c = \dfrac{[NH_3]^2}{[N_2]\,[H_2]^3} = \dfrac{(0.142\,mol\,dm^{-3})^2}{(1.36\,mol\,dm^{-3})\,(1.84\,mol\,dm^{-3})^3}$

$= 0.00238\,mol^{-2}\,dm^6$ *[2]*

2 $[C_2H_5OH] = \dfrac{[CH_3COOC_2H_5]\,[H_2O]}{[CH_3COOH]\,K_c}$

$= \dfrac{(3.0\,mol\,dm^{-3})\,(3.0\,mol\,dm^{-3})}{(0.80\,mol\,dm^{-3}) \times 4.0}$ *[1]*

$= 2.8125\,mol\,dm^{-3}$. *[1]*

7.4

1 a Increased concentration of reactants (C_2H_4 and H_2O) *[1]*

b Lower temperature *[1]*

c Higher pressure. *[1]*

2 a The blood-red colour darkens as the position of equilibrium moves right. *[1]*

b The blood-red colour becomes paler as the position of equilibrium moves left. *[1]*

3 a Increasing temperature moves the position of equilibrium left; increasing pressure moves the position of equilibrium right. *[2]*

b Increasing temperature moves the position of equilibrium right; increasing pressure has no effect on the position of equilibrium. *[2]*

8.1

1 a $0.05\,dm^3$ *[1]*

b $2.5\,dm^3$ *[1]*

2 a $2\,mol\,dm^{-3}$ *[1]*

b $0.05\,mol\,dm^{-3}$ *[1]*

3 a $4.8\,g$ *[1]*

b $6.3\,g$ *[1]*

4 a $0.02\,dm^3$ *[1]*

b $0.02 \times 0.100 = 0.002\,moles$ *[1]*

c $0.004\,moles$ *[1]*

d $0.004 \times \dfrac{1000}{25} = 0.16\,moles$ *[1]*

9.1

1 a K $+1$, Br -1

b H $+1$, O -2

c C $+2$, O -2

d P $+5$, O -2

e Mn $+4$, O -2

f Cr $+6$, O -2 *[6]*

2 a $CuCl_2$ **b** Cu_2O

c $PbCl_4$ **d** MnO_4^- *[4]*

3 $2Ca \rightarrow 2Ca^{2+} + 4e^-$ and $O_2 + 4e^- \rightarrow 2\,O^{2-}$. Calcium is oxidised and oxygen is reduced. *[4]*

4 $2Br^- + 2H^+ + H_2SO_4 \rightarrow Br_2 + SO_2 + 2H_2O$. Br is oxidised from -1 to 0; S is reduced from $+6$ to $+4$. *[2]*

9.2

1 a potassium at the cathode, chlorine at the anode

b calcium at the cathode, oxygen at the anode

c magnesium at the cathode, iodine at the anode. *[3]*

2 a hydrogen at the cathode, chlorine at the anode

b hydrogen at the cathode, oxygen at the anode

c copper at the cathode, oxygen at the anode. *[3]*

3 a $Al^{3+} + 3e^- \rightarrow Al$; $2Cl^- \rightarrow Cl_2 + 2e^-$

b $Pb^{2+} + 2e^- \rightarrow Pb$; $2H_2O(l) \rightarrow O_2(g) + 4H^+(aq) + 4e^-$

c $2H_2O(l) + 2e^- \rightarrow 2OH^-(aq) + H_2(g)$; $2Cl^- \rightarrow Cl_2 + 2e^-$ *[3]*

10.1

1 Increase the temperature, increase the concentration of sulfuric acid, divide the magnesium more finely (into a powder). *[1 mark each]*

2 Measure the rate at which hydrogen gas is produced, measure mass changes, measure pH changes as sulfuric acid is used up. *[1 mark each]*

3 As the reaction proceeds, reactants get used up. Therefore the concentration of reactants is greatest at the beginning. *[1]* The higher the concentration the faster the reaction because there are more frequent collisions between particles. *[1]*

10.2

1 The activation enthalpy is the minimum energy required by a pair of colliding particles before a reaction will occur. *[1]*

2

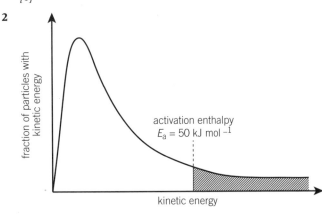

[2]

3 a The activation energy for the reaction between N_2 and O_2 is very high. However, inside a car engine the temperature is high enough for some molecules of N_2 and O_2 to collide with energies greater than the activation energy for the reaction, forming NO. *[1]*

b The activation energy for the reaction between NO and O_2 is low and many molecules have sufficient collision energy to exceed the activation energy at room temperature. *[1]*

10.3

1 a Manganese(IV) oxide is a catalyst for the reaction. *[1]*

b i manganese(IV) oxide is a heterogeneous catalyst *[1]*

ii catalase is a homogeneous catalyst. *[1]*

2 An intermediate is formed in the reaction. *[1]* This then reacts to form the product. *[1]*

3 The value of ΔH does not change if a catalyst is used. *[1]* The catalyst only affects the activation enthalpy. *[1]*

10.4

1 a) and c) are heterogeneous catalysts as they are in different states to the reactants. *[2]*

b) and d) are not heterogeneous catalysts as they are in the same state as the reactants. *[2]*

2 The reactant molecules are adsorbed onto the surface of the catalyst; bonds in the reactant break; new bonds form; the products are released from the catalyst surface and diffuse away. *[1 mark for each point]*

3 Lead is a catalyst poison for the metals in the catalytic converter. *[1]*

11.1

1 Around $-200\,°C$. The actual melting point of fluorine is $-219\,°C$. *[1]*

2 a fluorine   **b** calcium   **c** zinc *[3]*

3 a p-block   **b** s-block   **c** d-block *[3]*

4 a The third ionisation enthalpy of Mg is much higher than the second, so it is hard to remove the third electron and Mg forms Mg^{2+} *[1]*. For Al, it is the fourth ionisation enthalpy which is much higher so it is hard to remove the fourth electron, meaning Al forms Al^{3+} *[1]*.

b The total energy to form Mg^{2+} is $2\,186\,kJ\,mol^{-1}$ *[1]*. The total energy to form Al^{3+} is $5\,137\,kJ\,mol^{-1}$ *[1]*. It takes less energy for Mg to form ionic compounds so Mg is more reactive. *[1]*

11.2

1 $Sr(s) + 2H_2O(l) \rightarrow Sr(OH)_2(s) + H_2(g)$ *[1]*

2 $CaCO_3(s) \rightarrow CaO(s) + CO_2(g)$ *[1]*

3 a $Ba(OH)_2$ is more soluble *[1]*

b $MgCO_3$ is more soluble. *[1]*

4 $MgO(s) + 2HCl(aq) \rightarrow MgCl_2(aq) + H_2O(l)$ *[1]*

$Mg(OH)_2(s) + H_2SO_4(aq) \rightarrow MgSO_4(aq) + H_2O(l)$ *[1]*

11.3

1 $I_2(s)$ is a grey solid; $I_2(g)$ is a purple vapour; $I_2(aq)$ is a brown solution; I_2 in organic solvents is a purple solution. *[1 mark each]*

2 $Ag^+(aq) + I^-(aq) \rightarrow AgI(s)$ *[1]*; yellow precipitate. *[1]*

3 a $Cl_2(aq) + 2Br^-(aq) \rightarrow Br_2(aq) + 2Cl^-(aq)$ *[1]*

b brown colour *[1]*; bromine produced. *[1]*

c brown layer above a yellow aqueous layer *[1]*; bromine dissolves in the cyclohexane better than in water. *[1]*

d bromine is a weaker oxidising agent than chlorine. *[1]*

12.1

1 3-methylhexane has a chain of six carbons with a methyl group coming off the third carbon. 2,2,4-trimethylheptane has a chain of seven carbons with two methyl groups on the second carbon and one methyl group on the fourth. *[2]*

2 i $C_{10}H_{22}$

   ii $C_{10}H_{20}$ *[2]*

3 They all have the same molecular formula – C_6H_{12}. *[1]*

12.2

1 Pent-1-ene has the structure $CH_2=CHCH_2CH_2CH_3$. *[3]*

2 The product is pentane. *[1]*

3 H–Br has a dipole with Br having a partial negative charge. *[1]*

The C=C bond attacks the H–Br, causing heterolytic fission. *[1]*

A carbocation and a Br^- ion are formed. *[1]*

A lone pair on the Br^- ion attacks the carbocation forming the product 2-bromobutane. *[1]*

4 The polymer has the structure –CHCl–CHCl–CHCl–CHCl–CHCl–CHCl– *[2]*

12.3

1 The isomers are 1,2-dibromoethane and 1,1-dibromoethane. *[2]*

2 2-methylpropan-2-ol is $CH_3C(OH)(CH_3)CH_3$. 2-methylpropan-1-ol is $CH_3CH(CH_3)CH_2OH$. They are isomers. *[3]*

3 The carbons have two methyl groups or two hydrogen atoms attached to them. It does not have different groups attached to each carbon of the double bond. *[1]*

13.1

1 The differences are:

a 2-methylpentane has a CH_3 group attached to the second carbon of five in the chain, whereas 3-methylpentane has a CH_3 group attached to the third carbon. *[1]*

b 2,2-dimethylbutane has two CH_3 groups attached to the second carbon whereas 2,3-dimethylbutane has one CH_3 group on each of the second and third carbons. *[1]*

c 3-methylhexane has a six-carbon backbone whilst 3-methylheptane has a seven-carbon backbone. Both have a CH_3 group attached to the third carbon. *[1]*

2 They are incorrectly named because:

a 1-methylpropane has four carbons in the longest chain and should be called butane. *[1]*

b 2-ethylbutane has five carbons in the longest chain and should be called 3-methylpentane. *[1]*

c 2,3-methylbutane should be called 2,3-*di*methylbutane to highlight the presence of two methyl groups. *[1]*

3 Primary: butan-1-ol, secondary: butan-2-ol, tertiary: 2-methylpropan-2-ol. *[3 marks]*

4 The correct names are:

a 1-bromo-2-chloroethane as bromo is alphabetically before chloro. *[1]*

b 2-chloro-2-iodopropane as chloro is alphabetically before iodo. *[1]*

c 1,2,3-triiodopropane as '1,2,3-iodopropane' does not highlight the presence of three iodine atoms, and 'triiodopropane' does not give the position of the iodine atoms in the molecule. *[1]*

13.2

1 4-bromo-3,3-dichloro-2-iodohexane – the smallest possible numbers are chosen and the halogens are listed alphabetically. *[1]*

2 The following molecules have the higher boiling point.

a CH_3Br *[1]*; Br is bigger than F and has more electrons, so there are stronger instantaneous dipole–induced dipole bonds *[1]*

b CBr_4 *[1]*; it has more Br atoms in the molecule, more electrons, and therefore stronger instantaneous dipole–induced dipole bonds. *[1]*

3 The C–I bond is weaker than the C–F bond, so it is easier to break. *[1]*.

4 A nucleophile has a lone pair of electrons which it can use to form a covalent bond *[1]*; examples are H_2O, NH_3, OH^-, CN^-. *[1 mark each for any two]*

5 Heat the haloalkane in a sealed tube with concentrated ammonia solution. *[1]*

6 The mechanism is similar to that between hydroxide ions and 1-bromobutane shown, except that the haloalkane involved is $CH_3CH_2CH_2I$. *[1 mark for showing the partial charges on the C-I bond; 1 mark for correct arrow from the nucleophile to the partially positive carbon atom; 1 mark for showing the C–I bond breaking; 1 mark for the correct product (propan-1-ol).*

13.3

1 The homologous series are:

a aldehyde *[1]*

b ketone *[1]*

c carboxylic acid *[1]*

d (secondary) alcohol *[1]*

e ester. *[1]*.

2 Butanal *[1]* and butanoic acid. *[1]*

3 The colour changes are:

    a No change – remains orange *[1]* as 2-methylpropan-2-ol is a tertiary alcohol and is not oxidised. *[1]*

    b Orange to green *[1]* as 2-methylpropan-1-ol is a primary alcohol *[1]* and is oxidised, thereby reducing the dichromate ion to green Cr^{3+}. *[1]*

4 Acidified *[1]* potassium dichromate(VI) *[1]* with distillation. *[1]*

13.4

1 The names are:

    a hexanoic acid *[1]*

    b pentanedioic acid *[1]*

    c 3-methylbutanoic acid. *[1]*

2 $CH_3CH_2COOH + KOH \rightarrow CH_3CH_2COONa + H_2O$, *[1 mark for each product]*

3 Propan-1-ol *[1]*, CH_3CH_2COOH *[1]* and butanoic acid *[1]*, $CH_3CH_2CH_2OH$. *[1]*

4 The samples which would react are:

    a 4-methylphenol *[1]*

    b ethanoic acid *[1]*

    c propan-1-ol *[1]*

    d 4-methylphenol *and* ethanoic acid. *[1]*

5 Phenols are not acidic enough to react with sodium carbonate solution. *[1]*

14.1

1 A batch process produces small quantities at a time whilst a continuous process produces large quantities and runs without interruption. *[1]*

2 **a** continuous

    b batch

    c continuous. *[3]*

3 There may already be good transport links / could use products from the existing site / cheap energy supply / good water supply / skilled labour / may be near existing raw materials – any three suggestions. *[3]*

14.2

1 **a** % atom economy

$$= \frac{\text{relative formula mass of useful product}}{\text{relative formula mass of reactions used}} \times 100$$

$$= \frac{2 \times 44}{(2 \times 28) + 32} \times 100 = 100\% \qquad [1]$$

    b It is reused by separating it from the product and putting it back in at the start of the plant. *[1]*

2 Generally addition > substitution > elimination. *[1]* Addition reactions have atom economies of 100%; elimination reactions have low atom economies as there is always a waste product; substitution reactions tend to be in the middle depending on which atoms are removed from the reactant. If a heavy atom such as Br is substituted and the Br-containing product is a waste product, the atom economy could be lower than for the elimination reaction. *[1]*

3 atom economy for step 1 $= \dfrac{99}{28 + 71} \times 100 = 100\%$ *[1]*

    atom economy for step 2 $= \dfrac{62.5}{99} \times 100 = 63.1\%$ *[1]*

15.1

1 $N_2 + O_2 \rightarrow 2NO$ *[1]*. NO is oxidised to NO_2 which dissolves in rain water to make acid rain. *[1]*

2 Carbon monoxide is toxic to humans because it binds irreversibly to haemoglobin, reducing the uptake of oxygen by the blood. *[1]* A possible equation is $2CH_4 + 3O_2 \rightarrow 2CO + 4H_2O$ *[1]*, but other balanced equations forming CO and H_2O (possibly together with CO_2 and/or C) could be credited.

3 There is more complete combustion. *[1]*

15.2

1 Hydrogen has a low density and may easily escape. It is highly flammable. *[1]*

2 Advantage: renewable source, carbon neutral. Disadvantage: may use up farmland which could be used for food; may promote deforestation. *[2]*

3 It is made by fermenting sugar. Sugar cane only grows in certain areas. *[1]*

Answers to practice questions

Chapter 1

1 B *[1]*, **2** D *[1]*, **3** C *[1]*, **4** C *[1]*, **5** D *[1]*

6 a Mol NaOH = (15.75/1000) × 0.500
= $7.875 × 10^{-3}$ mol *[1]*

b Mol HCl = mol NaOH (because 1:1 ratio in the equation) = $7.875 × 10^{-3}$ mol. *[1]*

c Concentration of HCl = moles/(vol/1000)
= $7.875 × 10^{-3}$/(10.0/1000) = 0.7875 *[1]*
= 0.788 mol dm^{-3} to 3 s.f. *[1]*

d Concentration HCl in undiluted solution
= 0.788 × dilution factor = 7.88. *[1]*

Dilution factor is 10, because 25 cm^3 was used to make up 250 cm^3 of diluted solution *[1]*
= 0.788 × 10 = 7.88 mol dm^{-3}.

Mass of 7.88 mol = 7.88 × M_r(HCl)
= 7.88 × 36.5 = 287.62.

So concentration in g dm^{-3} = 288
(to 3 or more s.f.). *[1]*

e Take 50 cm^3 of 2.00 mol dm^{-3} NaOH. *[1]*

Measure out using a 50 cm^3 volumetric pipette / 2 × 25 cm^3 volumetric pipette. *[1]*

Add water up to the mark. *[1]*

Fine detail: ONE of: add water dropwise when close to mark, invert solution several times to ensure mixing, observe level of liquid in pipette or flask at eye level. *[1]*

7 a Test tube connected a tube to a gas syringe OR inverted burette / measuring cylinder in trough of water *[1]*. Magnesium carbonate labelled AND heat shown being applied to magnesium carbonate *[1]*, apparatus shown as airtight (for gas syringe), OR delivery tube shown in appropriate place under burette. *[1]*

b 2.00 g MgCO$_3$ = 2.00/84.3 = 0.082 30 mol *[1]*. Volume of CO$_2$ formed = 0.082 30 × 24.0 = 1.98 dm^3 *[1]*. Answer given to 2 or more s.f.

c $v = nRT/p$ *[1]* = 1 × 8.314 × 210/600 = 2.91 (m^3)
[1] = 2910 dm^3. *[1]*

Chapter 2

1 A *[1]*, **2** C *[1]*, **3** B *[1]*, **4** D *[1]*

5 a i **Two or more** (lighter) nuclei **join** together, to form a **single heavier** nucleus. *[2]*

ii To overcome the repulsion *[1]* between the **positive** nuclei. *[1]*

b i $^1_1H + ^2_1H \rightarrow ^3_2He$ *[1 for each H isotope]*

ii $^{12}_6C$ *[1: numbers, 1: symbol]*

iii The Sun is not hot enough / does not have high enough pressure (at centre) / is not a heavyweight star.

OR

Other stars are hotter / have a higher pressure (at centre) / are heavyweight stars. ALLOW that there is no hydrogen left in these stars. *[1]*

6 a i 2 electrons in sub-shell A, 2 electrons in sub-shell B.

[1 mark: 4 electrons in total, 1 mark: electrons in correct sub-shells]

Electrons shown with opposite spins in sub-shell A, AND electrons in separate orbitals in sub-shell B (ignore direction of spins). *[1]*

ii 10 *[1]*

iii A = 3s, B = 3p

[2 marks, ALLOW 1 mark if s and p both correct but numbers are wrong]

iv s-orbitals are spherical, p-orbitals are dumbbell shaped. ALLOW suitable sketch. *[1]*

OR p-orbitals have two regions of electron density, s-orbitals only have one.

OR p-orbitals have no electron density close to centre of atom.

b

Species	Electron configuration
Na$^+$	1s^{2}2s^{2}2p^6 IGNORE 3s^0 *[1]*
F, ALLOW O$^-$ Ne$^+$ etc. *[1]*	1s^{2}2s^{2}2p^5
Cr	[Ar] 4s^2 *[1]* 3d^4 *[1]* in either order

[4]

c i p-orbitals / sub-shells have the highest energy OR are the outermost electrons. *[1]*

ii 5p^3. *[1]*

Chapter 3

1 B *[1]*, **2** D *[1]*, **3** B *[1]*,

4 C *[1]*, **5** A *[1]*

6 a i NH$_4^+$ and Cl$^-$ ions shown alternating in rows *[1]* at least 2 rows shown, with ions of opposite charges shown as nearest neighbours throughout. *[1]*

ii There is electrostatic attraction *[1]* between + and – charged ions. *[1]*

b i One of the bonds has both electrons from the same atom *[1]* both come from the N atom. *[1]*

ii Electron pairs / groups repel each other. *[1]*

To get as far apart as possible / to minimise repulsion. *[1]*

4 groups / pairs of electrons around N atom. *[1]*

Tetrahedral arrangement (of electrons and hence tetrahedral shape of molecule). *[1]*

Bond angle is 109½°. *[1]*

c   Triple bond between N atoms *[1]*. Correct number of electrons on each N atom AND lone pair shown on each N atom. *[1]*

Chapter 4

1 A *[1]*,   2 C *[1]*,   3 B *[1]*,   4 C *[1]*,   5 B *[1]*

6 a   E = m c ΔT = 100 × 4.18 × 24 = 10032 (J) *[1]*, correct values for m and ΔT *[1]*. Correct evaluation of E (with e.c.f.).

   b i   Moles of ethanol = 1.10/46 = 0.02391 *[1]* enthalpy change per mole =10032/0.02391 = 418600 J OR 418.6 kJ *[1]*. $\Delta_c H = -419 \, kJ \, mol^{-1}$ *[2]*

      ii   Heat loss (to the surroundings) / incomplete combustion of ethanol / evaporation of ethanol during the experiment. *[ANY 2]*

7 a i   Bonds broken = 3C × C + 8C × H + 1C=O + 6 O=O *[1]*. Bonds formed = 8 C=O + 8O−H *[1]*. Energy required to break bonds = 8138, energy released by forming bonds = (−)10152 *[1]*. Enthalpy of combustion = 8138 − 10152 = −2104 kJ mol⁻¹. *[1]* mistake = −2014 kJ/mol

      ii   Bond enthalpies are averages / are not the same as the actual values in the molecule. *[1]* Bond enthalpies can only be used for molecules in gaseous state ALLOW butanone and water are in the liquid state under standard conditions. *[1]*

   b   (Enthalpy change will be similar) *[1]* because the number and type of bonds in butanone are the same as butanal *[1]* so the same (number and type of) bonds are broken and formed when both molecules react. *[1]*

Chapter 5

1 C *[1]*,   2 B *[1]*,   3 B *[1]*,   4 C *[1]*,
5 D *[1]*,   6 A *[1]*,   7 B *[1]*

8 a i   Positive and negative charges at opposite ends of the bond / uneven distribution of charge in the bond. *[1]*

      ii   F and Cl are both more electronegative *[1]* than C. *[1]*

      iii   F has a greater negative charge than the Cl atoms *[1]*; centre of negative charge is not in the centre of the molecule / is not in the same place as the centre of positive charge / charges are arranged unsymmetrically. *[1]*

      iv   Permanent dipole–permanent dipole bonds. *[1]*

   b i   Random movement of electrons *[1]*; results in an unequal distribution of charge *[1]*. This induces a dipole on a neighbouring molecule *[1]*; the two dipoles /molecules attract each other. *[1]*

      ii CCl_4 has more electrons than CCl_3F *[1]* so more change of an instantaneous dipole arising *[1]* instantaneous dipole–induced dipole bonds are stronger *[1]* (more significant than) additional permanent dipole–permanent dipole bonds between CCl_3F molecules *[1]* overall, more energy needed to separate CCl_4 molecules. *[1]*

Chapter 6

1 D *[1]*,   2 A *[1]*,   3 A *[1]*,   4 D *[1]*,   5 D *[1]*

6 a i   Ozone absorbs high frequency / high energy ultraviolet radiation *[1]* this causes skin cancer/ eye cataracts (if it reaches the Earth's surface). *[1]*

      ii   CFC molecules diffuse from troposphere to stratosphere / are not broken down in the troposphere *[1]* in the stratosphere, high frequency / energy uv light breaks the C−Cl bond *[1]* Cl radicals are formed *[1]* these act as a catalyst for the break down of ozone. *[1]*

   b i   Homolytic *[1]*

      ii   A Species with an unpaired electron. *[1]*

      iii   Energy required to break one bond = 305 × 1000 *[1]* / 6.02 × 1023 = 5.07 × 10⁻¹⁹ J *[1]*

      iv $f = \dfrac{E}{h}$ *[1]* $= \dfrac{5.07 \times 10^{-19}}{6.63 \times 10^{-34}} = 7.65 \times 10^{14} \, Hz$ *[1]*

      v   O atom combines with O_2 to form O_3. *[1]*

7 a i   Bonds vibrate *[1]* with more energy *[1]*

      ii   Peak at 2510–3490 indicates O−H in carboxylic acid *[1]* no O−H in propanal *[1]* peak at 1715 indicates C=O in carboxylic acid *[1]* C=O in aldehyde would be seen at 1720–1740. *[1]*

   b i   Peak at 74 is molecular ion / unfragmented propanoic acid *[1]*. Peak at 75 is due to presence of a ¹³C atom. *[1]*

      ii   Peak at 57 = $C_3H_5O^+$ (loss of OH) *[1]*. Peak at 45 = COOH⁺. *[1]* (1 max if no + charges on either species).

Chapter 7

1 C *[1]*,   2 D *[1]*

3 a   pH 8–10 due to the presence of OH⁻ ions. *[1]*

   b   H⁺ ions from the acid will react with OH⁻ ions, removing them from the system *[1]*; the position of equilibrium will move to the right to replace the OH⁻ ions. *[1]*

4 a i Step 1: increasing temperature moves position of equilibrium to the right as the forward reaction is endothermic [1]; increasing pressure moves position of equilibrium to the left as there are fewer molecules on the left [1]; catalyst does not affect the position of equilibrium. [1]

Step 2: increasing temperature moves position of equilibrium to the left as the forward reaction is exothermic [1]; increasing pressure moves position of equilibrium to the right as there are fewer molecules on the right [1]; catalyst does not affect the position of equilibrium. [1]

ii Step 2: is exothermic [1] so a high temperature moves the position of equilibrium to the left, making less product [1]; in step 1 a higher temperature moves the position of equilibrium to the right, making more product. [1]

iii Step 2: has fewer molecules on the right than on the left [1] so a higher pressure moves the position of equilibrium to the right, making more product [1]; in step 1 a higher temperature moves the position of equilibrium to the left, making less product. [1]

b i $K_c = \dfrac{[CH_3OCH_3][H_2O]}{[CH_3OH]^2}$ [2]

ii $K_c = \dfrac{(0.20) \times (0.20)}{(0.050)^2} = 16$

[1 mark for correct numbers in expression; 1 mark for identifying there are no units for K_c in this expression.]

Chapter 8

1 B [1], **2** A [1], **3** C [1], **4** C [1]

5 a 2.5×10^{-4} mol [1]

b 2.5×10^{-4} mol [1]

c 0.0127 mol dm^{-3} [1]

6 a 0.005 mol [1]

b 0.01 mol [1]

c 0.355 mol dm^{-3} [1]

7 a i $\dfrac{25.4 \times 0.1}{1000}$ [1] $= 0.00254$ [1]

ii $0.00254 \times 0.5 = 0.00127$ [1]

iii 0.00127 [1]

iv $\dfrac{0.00127 \times 1000}{5}$ [1] $= 0.254$ mol dm^{-3} [2]

b 0.254×51.5 [1] $= 13.1$ g dm^{-3}. [2]

Chapter 9

1 A [1], **2** B [1], **3** C [1],

4 C [1], **5** D [1], **6** B [1]

7 $2\,I^- + 4\,H^+ + MnO_2 \rightarrow I_2 + 2\,H_2O + Mn^{2+}$ [2]

8 Nitrogen is oxidised; iodine is reduced. [2]

9 a From colourless [1] to red/brown. [1]

b Loss of electrons / increase in oxidation state. [1]

c $2Cl^- \rightarrow Cl_2 + 2e^-$ [1]

d $Cl_2 + 2Br^- \rightarrow Br_2 + 2Cl^-$ [1]

e Chlorine oxidises the bromide ions [1]; chlorine is reduced. [1]

Chapter 10

1 C [1], **2** A [1], **3** A [1], **4** D [1], **5** D [1]

6 a Heterogeneous [1]

b More particles will have energy greater than the activation enthalpy [1]; particles collide more frequently and the reaction is faster. [1]

c It has a greater surface area [1] so collisions are more frequent. [1]

d It provides an alternative pathway [1] of lower activation enthalpy. [1]

e Reactants are adsorbed onto the catalyst [1]; bonds in the reactant break [1]; new bonds form [1]; products detach from the catalyst surface. [1]

7 a Homogeneous as it is in the same state as the reactant. [1]

b NO forms an intermediate with ozone [1] and provides an alternative pathway [1] of lower activation enthalpy. [1]

c Higher temperature means particles move faster/collide with greater energy. [1] Higher proportion [1] of molecules collide with energy greater than the activation enthalpy. [1]

Chapter 11

1 D [1], **2** D [1], **3** D [1], **4** A [1],

5 A [1], **6** D [1], **7** C [1], **8** D [1],

9 A [1], **10** C [1]

11 a Bromine vapour is toxic; it spreads far because it is volatile; bromine is corrosive. [Max 2 marks]

b Instantaneous dipole–induced dipole bonds [1] are stronger between bromine molecules [1] as the M_r is greater/larger number of electrons [1] more energy is needed to break them (so a higher b.p.) [1]

12 Mg^{2+} and O^{2-} ions have high charge densities [1]. They attract each other strongly and it takes a lot of energy to separate them [1].

13 a $CaCO_3 \rightarrow CaO + CO_2$ *[1]*

  b $MgCO_3$ *[1]* because Mg^{2+} is a smaller ion which polarises the carbonate ion more. *[1]*

  c Calcium oxide *[1]*

  d $CaO + 2HA$ *[1]* $\rightarrow CaA_2 + H_2O$ *[1]*

Chapter 12

1 C *[1]*, **2** A *[1]*, **3** C *[1]*, **4** B *[1]*, **5** C *[1]*

6 It must be an alkene as it reacts with bromine and produces an addition polymer *[1]*. It must have four carbon atoms. *[1]* It must have different groups on each carbon of the C=C bond as it has E/Z isomers. *[1]* It is but-2-ene. *[1]*

7 a $C_{10}H_{22}$ *[1]*

  b There are four groups of electrons around each carbon atom *[1]* which repel *[1]* as far as possible *[1]*.

  c C_3H_6 is propene and C_7H_{14} is heptane. *[2]*

  d C_3H_6 has no isomers. C_7H_{14} can show chain isomerism. *[2]*

  e Electrophilic addition. *[1]*

  f 1,2-dibromopropane. *[1]*

  g The lone pair on the oxygen of the water molecule can attack the carbocation to produce $CH_2BrCH(OH)CH_3$. *[1]*

Chapter 13

1 C *[1]*, **2** D *[1]*, **3** A *[1]*, **4** D *[1]*,

5 B *[1]*, **6** D *[1]*, **7** A *[1]*, **8** D *[1]*

9 The observations are:

  a Violet colour. *[1]*

  b No reaction. *[1]*

  c No reaction. Phenols need acid anhydrides for esterification. *[1]*

10 a Orange. *[1]*

  b It is a tertiary alcohol *[1]* as it has not been oxidised. *[1]*

  c 2-methylpropan-2-ol [1]; $CH_3C(CH_3)(OH)CH_3$. *[1]*

  d $CH_3CH_2CH_2CH_2OH$ (butan-1-ol) *[1]* and $CH_3CH(CH_3)CH_2OH$ (2-methyl-propan-1-ol). *[1]*

  e Oxidation of butan-1-ol would give butanal, $CH_3CH_2CH_2CHO$ *[1]* under mild conditions, and butanoic acid, $CH_3CH_2CH_2COOH$ *[1]* under reflux. Oxidation of 2-methyl-propan-1-ol would give 2-methylpropanal, $CH_3CH(CH_3)CHO$ *[1]* under mild conditions, and 2-methylpropanoic acid, $CH_3CH(CH_3)COOH$ *[1]* under reflux.

11 a Carbon has a partial positive charge and the halogen has a partial negative charge *[1]* because the halogen is more electronegative. *[1]*

  b Boiling point. *[1]*

  c The nucleophile is attracted to the partial positive charge of the carbon atom; *[1]* a bond forms between the nucleophile and the carbon atom; *[1]* the C-Hal bond breaks. *[1]*

Chapter 14

1 C *[1]*, **2** D *[1]*, **3** A *[1]*, **4** D *[1]*

5 a Acidified *[1]* dichromate *[1]* reflux *[1]*

  b Propanone: ketone *[1]*; propan-2-ol: (secondary) alcohol *[1]*

  c 96.7% *[1]*

  d 0.1 mol *[1]*

  e 0.1 mol *[1]*

  f 5.8 g *[1]*

  g 45% *[1]*

6 a 42.2% *[1]*

  b Amount of $C_2H_5Br = 0.05$ mol *[1]*; expected amount of $C_2H_5OH = 0.05$ mol *[1]*; expected mass of $C_2H_5OH = 2.30$ g *[1]*; percentage yield = 62%. *[1]*

Chapter 15

1 C *[1]*, **2** D *[1]*, **3** A *[1]*,

4 B *[1]*, **5** D *[1]*, **6** B *[1]*

7 a $C_{10}H_{22} + 15\frac{1}{2} O_2 \rightarrow 10CO_2 + 11H_2O$ *[1]*

  b $NO + \frac{1}{2}O_2 \rightarrow NO_2$ *[1]*

  c $SO_2 + \frac{1}{2}O_2 \rightarrow SO_3$ *[1]*

  d $SO_3 + H_2O \rightarrow H_2SO_4$ *[1]*

8 It is made from plants *[1]* which can be regrown *[1]*; it is carbon neutral *[1]*; it produces less CO, C_xH_y, SO_2, and particulates than from a diesel engine *[1]*; it is biodegradable. *[1]* *[max 4 marks]*

For Reference

List of constants

Molar gas volume $= 24.0\,dm^3\,mol^{-1}$ at RTP

Avogadro constant, $N_A = 6.02 \times 10^{23}\,mol^{-1}$

Specific heat capacity of water, $c = 4.18\,J\,g^{-1}\,K^{-1}$

Planck constant, $h = 6.63 \times 10^{-34}\,J\,Hz^{-1}$

Speed of light in a vacuum, $c = 3.00 \times 10^8\,m\,s^{-1}$

Ionic product of water, $K_w = 1.00 \times 10^{-14}\,mol^2\,dm^{-6}$ at 298 K

1 tonne $= 10^6\,g$

Arrhenius equation: $k = Ae^{-E_a/RT}$ or $\ln k = -E_a/RT + \ln A$

Gas constant, $R = 8.314\,J\,mol^{-1}\,K^{-1}$

Characteristic infrared absorptions in organic molecules

Bond	Location	Wavenumber / cm^{-1}
C—H	Alkanes	2850–2950
	Alkenes, arenes	3000–3100
C—C	Alkanes	750–1100
C=C	Alkenes	1620–1680
aromatic C=C	Arenes	Several peaks in range 1450–1650 (variable)
C=O	Aldehydes	1720–1740
	Ketones	1705–1725
	Carboxylic acids	1700–1725
	Esters	1735–1750
	Amides	1630–1700
	Acyl chlorides and acid anhydrides	1750–1820
C—O	Alcohols, ethers, esters and carboxylic acids	1000–1300
C≡N	Nitriles	2220–2260
C—X	Fluoroalkanes	1000–1350
	Chloroalkanes	600–800
	Bromoalkanes	500–600
O—H	Alcohols. phenols	3200–3600 (broad)
	Carboxylic acids	2500–3300 (broad)
N—H	Primary amines	3300–3500
	Amides	*ca.* 3500

The Periodic Table of Elements

Key

atomic number
Symbol
name
relative atomic mass

(1)	(2)												(3)	(4)	(5)	(6)	(7)	(0)
1																		**18**
1 **H** hydrogen 1.0	**2**												**13**	**14**	**15**	**16**	**17**	2 **He** helium 4.0
3 **Li** lithium 6.9	4 **Be** beryllium 9.0												5 **B** boron 10.8	6 **C** carbon 12.0	7 **N** nitrogen 14.0	8 **O** oxygen 16.0	9 **F** fluorine 19.0	10 **Ne** neon 20.2
11 **Na** sodium 23.0	12 **Mg** magnesium 24.3	**3**	**4**	**5**	**6**	**7**	**8**	**9**	**10**	**11**	**12**		13 **Al** aluminium 27.0	14 **Si** silicon 28.1	15 **P** phosphorus 31.0	16 **S** sulfur 32.1	17 **Cl** chlorine 35.5	18 **Ar** argon 39.9
19 **K** potassium 39.1	20 **Ca** calcium 40.1	21 **Sc** scandium 45.0	22 **Ti** titanium 47.9	23 **V** vanadium 50.9	24 **Cr** chromium 52.0	25 **Mn** manganese 54.9	26 **Fe** iron 55.8	27 **Co** cobalt 58.9	28 **Ni** nickel 58.7	29 **Cu** copper 63.5	30 **Zn** zinc 65.4		31 **Ga** gallium 69.7	32 **Ge** germanium 72.6	33 **As** arsenic 74.9	34 **Se** selenium 79.0	35 **Br** bromine 79.9	36 **Kr** krypton 83.8
37 **Rb** rubidium 85.5	38 **Sr** strontium 87.6	39 **Y** yttrium 88.9	40 **Zr** zirconium 91.2	41 **Nb** niobium 92.9	42 **Mo** molybdenum 95.9	43 **Tc** technetium	44 **Ru** ruthenium 101.1	45 **Rh** rhodium 102.9	46 **Pd** palladium 106.4	47 **Ag** silver 107.9	48 **Cd** cadmium 112.4		49 **In** indium 114.8	50 **Sn** tin 118.7	51 **Sb** antimony 121.8	52 **Te** tellurium 127.6	53 **I** iodine 126.9	54 **Xe** xenon 131.3
55 **Cs** caesium 132.9	56 **Ba** barium 137.3	57–71 lanthanoids	72 **Hf** hafnium 178.5	73 **Ta** tantalum 180.9	74 **W** tungsten 183.8	75 **Re** rhenium 186.2	76 **Os** osmium 190.2	77 **Ir** iridium 192.2	78 **Pt** platinum 195.1	79 **Au** gold 197.0	80 **Hg** mercury 200.6		81 **Tl** thallium 204.4	82 **Pb** lead 207.2	83 **Bi** bismuth 209.0	84 **Po** polonium	85 **At** astatine	86 **Rn** radon
87 **Fr** francium	88 **Ra** radium	89–103 actinoids	104 **Rf** rutherfordium	105 **Db** dubnium	106 **Sg** seaborgium	107 **Bh** bohrium	108 **Hs** hassium	109 **Mt** meitnerium	110 **Ds** darmstadtium	111 **Rg** roentgenium	112 **Cn** copernicium		113	114 **Fl** flerovium	115	116 **Lv** livermorium		

57 **La** lanthanum 138.9	58 **Ce** cerium 140.1	59 **Pr** praseodymium 140.9	60 **Nd** neodymium 144.2	61 **Pm** promethium 144.9	62 **Sm** samarium 150.4	63 **Eu** europium 152.0	64 **Gd** gadolinium 157.2	65 **Tb** terbium 158.9	66 **Dy** dysprosium 162.5	67 **Ho** holmium 164.9	68 **Er** erbium 167.3	69 **Tm** thulium 168.9	70 **Yb** ytterbium 173.0	71 **Lu** lutetium 175.0
89 **Ac** actinium	90 **Th** thorium 232.0	91 **Pa** protactinium	92 **U** uranium 238.1	93 **Np** neptunium	94 **Pu** plutonium	95 **Am** americium	96 **Cm** curium	97 **Bk** berkelium	98 **Cf** californium	99 **Es** einsteinium	100 **Fm** fermium	101 **Md** mendelevium	102 **No** nobelium	103 **Lr** lawrencium